CONTINGENCY TABLE ANALYSIS FOR ROAD SAFETY STUDIES

NATO ADVANCED STUDY INSTITUTES SERIES

Proceedings of the Advanced Study Institute Programme, which aims at the dissemination of advanced knowledge and the formation of contacts among scientists from different countries.

The series is published by an international board of publishers in conjunction with NATO Scientific Affairs Division

A	Life Sciences	Plenum Publishing Corporation
B	Physics	London and New York
C	Mathematical and Physical Sciences	D. Reidel Publishing Company Dordrecht and Boston
D	Behavioural and Social Sciences	Sijthoff & Noordhoff International Publishers B.V.
E	Applied Sciences	Alphen aan den Rijn, The Netherlands and Rockville, Md., U.S.A.

Series E: Applied Sciences - No. 42

CONTINGENCY TABLE ANALYSIS FOR ROAD SAFETY STUDIES

edited by

GERALD A. FLEISCHER

Professor of Industrial and Systems Engineering
University of Southern California,
Los Angeles, U.S.A

Springer-Science+Business Media, B.V.

Proceedings of the NATO Advanced Study Institute on
Contingency Table Analysis for Road Safety Studies
Sogesta, Italy
June 18-29, 1979

ISBN 978-94-009-8599-5   ISBN 978-94-009-8597-1 (eBook)
DOI 10.1007/978-94-009-8597-1

Copyright © 1981 Springer Science+Business Media Dordrecht
Originally published by Sijthoff & Noordhoff International Publishers B.V.,
Alphen aan den Rijn, The Netherlands 1981
Softcover reprint of the hardcover 1st edition 1981

All rights reserved. No part of this publication may be reproduced, stored in a retrieval system, or trans
mitted, in any form or by any means, electronic, mechanical, photocopying, recording, or otherwise
without the prior permission of the copyright owner.

CONTINGENCY TABLE ANALYSIS

FOR

ROAD SAFETY STUDIES

CONTENTS

FOREWORD

The analysis of statistical data is a critical
element in road safety studies. For example, specific
projects or programs may be implemented with the analyst
asked to answer the question, "What has been the effect
of this project (program) on accident frequency and/or
severity?" Are there any interdependencies or contribut-
ing effects due to the age, sex or driving experience of
involved motorists? What is the contribution, if any,
of roadway design, time of day, traffic density, etc.?"
To answer, or to provide insight into, these types of
questions, contingency tables are often used to display
frequency or count data. The subsequent analysis of
these contingency tables is the principal form of this
book.

Because of recent advances in the underlying
statistical methodology and procedures, and because
of the increasing interest in the application of
contingency table analysis to road safety studies, an
Advanced Study Institute (ASI) directed to this topic
was held at the Sogesta Conference Center, Urbino,
Italy, during the period 18-29 June 1979. The ASI
was funded by the Scientific North Atlantic Treaty
Organization (NATO) as part of its Advanced Study
Institutes Programme. The contents of this book,
with two exceptions described below, represent the
Proceedings of the ASI.

This book is organized into two principal sections.
Part I is a set of papers describing the underlying

theory and methodologies appropriate to contingency table analysis. In particular, it includes an extensive discussion of the minimum discrimination information (MDI) approach which leads to log-linear models and enables the analyst to find estimates of cell entries under various hypotheses or models and to test these hypotheses or models. Also included is a discussion of CONTAB, a PL/1 computer program which computes maximum liklihood estimates of parameters in the log-linear model for contingency tables and computes the associated test statistics. (The CONTAB material was originally prepared by John Nolan of the Department of Statistics, The George Washington University, Washington, D.C., USA. Mr. Nolan was not a participant at the ASI.)

Part II of the text includes a series of illustrations in which the methodologies discussed previously are applied to several road safety analyses conducted in Europe and the United States. One of these, Dr. James Hedlund's paper, "The Severity of Large Truck Accidents," describes a particularly interesting application of the MDI approach to examine the severity of accidents between cars and certain large trucks in the United States over a 2-year period. (Although Dr. Hedlund was not a participant in the ASI, his paper has been included in this volume because of its special relevance.) The applications in Part II are in addition to several other numerical examples included by the authors of certain of the methodological papers in Part I.

In closing, I want to express my appreciation to the international group of participants, lecturers and students alike, whose enthusiasm and intellectual effort contributed to the success of the ASI from which these Proceedings are derived. Their good-humored cooperation and collegiality is gratefully acknowledged. And we are all most appreciative of the understanding and support provided by the Directors of the Scientific Affairs Division of NATO, Dr. Tilo Kester (until the fall of 1978) and his successor, Dr. Mario d. Lullo.

G. A. Fleischer, Ph.D.
Los Angeles, California
November 1979

PART I

STATISTICAL MODELS AND METHODOLOGY

CONTINGENCY TABLES*

E.B. Andersen

Institute of Statistics
University of Copenhagen, Denmark

1. INTRODUCTION

The following terminology is convenient: A contingency table is formed by classifying a random sample from a given population into certain categories according to two (or more) criteria.

One of the reasons for collecting data in the form of contingency tables is to study relationships between different criteria. It is, thus, important if it can be shown that an individual belonging to a certain category according to criterion 1 is more likely to belong to certain categories according to criterion 2. Relationships of this sort are often described as interactions between two criteria.

2. SAMPLING DESIGNS

Contingency tables can be established in different ways, but several of the possible sampling designs lead to basically the same statistical model. We consider such a set of three possible sample designs and show how they are related.

---

*The material included in this paper is taken from Professor Anderson's new textbook currently in press. We are grateful to the publisher North Holland Publishing Company, for permission to reproduce this material here. The interested reader is referred to Professor Andersen's textbook for a more complete discussion of contingency tables.

Let the contingency table be a two way table of dimension m k. The observed number in cell (i, j) is denoted by x_{ij} and the corresponding random variable is X_{ij}. The quantity x_{ij} thus denotes the number of individuals from the population falling into category i according to criterion 1 and category j according to criterion 2. The probability model is different for each of the three designs.

Sampling Design I: Assume that X_{ij} is Poisson distributed with parameter λ_{ij} and that all X_{ij}'s are independent. This model can also be described as a contingency table with no marginals given. For sampling design I, the most important hypothesis is

(1) $H_o^I: \lambda_{ij} = \lambda_{i.} \lambda_{.j} / \lambda_{..}$,

where $\lambda_{i.} = \Sigma \lambda_{ij}, \lambda_{.j} = \Sigma \lambda_{ij}$ and $\lambda_{..} = \Sigma_i \Sigma_j \lambda_{ij}$.

Under H_o^I the model is called the <u>multiplicative</u> Poisson model, and we may call H_o^I the <u>multiplicativity hypothesis</u>.

Sampling Design II: Assume that all X_{ij}'s are independent and that $(X_{11}, \ldots, X_{mk})$ follow a multinomial distribution with parameter

$n = \sum_{i=1}^{m} \sum_{j=1}^{k} X_{ij}$,(which is fixed),

and certain cell probabilities $p_{11}, \ldots, p_{mk}$. Since the total n of the contingency table is a parameter, and hence not stochastic, we describe this case as a design with <u>fixed over-all total</u>. For sampling design II, the most important hypothesis is

(2) $H_o^{II}: p_{ij} = p_{i.} p_{.j}$,

where $p_{i.} = \sum_{j=1}^{k} p_{ij}$ and $p_{.j} = \sum_{i=1}^{m} p_{ij}$.

We note that $p_{i.}$ is the marginal probability that an individual falls in category i according to criterion 1 and that $p_{.j}$ is the marginal probability that an individual falls in category j according to criterion 2. Hence, (2) is a hypothesis of independence between events connected with criterion 1 and criterion 2. By the laws of probability, (2) is equivalent to

P{falling in cell i,j| in row i} = P {being in column j}.

Under H_o^{II}, the characterizations according to criteria 1 and 2 are accordingly independent. These considerations have led to the term independence hypothesis for H_o^{II}.

Sampling Design III: Let again all X_{ij}'s be independent, but assume that for each $i=1,\ldots,m$, we have that $(X_{i1},\ldots,X_{ik})$ is multinomially distributed with parameters

$$n_i = \sum_{j=1}^{k} X_{ij}, \text{ which is fixed,}$$

and cell probabilities $p_{i1}^*,\ldots,p_{ik}^*$. Since the row marginals are not stochastic, we describe this case as a design with fixed row marginals. The most important hypothesis for this design is

(3) H_o^{III}: $p_{ij}^* = p_{.j}^*/m$, where $p_{.j}^* = \sum_i p_{ij}^*$.

Since this hypothesis is equivalent to $p_{ij}^* = p_{lj}^*$ for any $l \neq i$, the m multinomial destributions of the design have identical cell probabilities under the hypothesis. This situation is often described as homogeneity of the row distributions and the hypothesis H_o^{III} is called the homogeneity hypothesis.

The following two theorems show that only a specification of the marginals we assume fixed, separates the three designs. In addition, hypotheses H_o^{I}, H_o^{II}, and H_o^{III} are equivalent.

Theorem 1*: Sampling design II follows from sampling design I by conditioning upon $X.. = n$, and H_o^{I} transforms into H_o^{II}.

Theorem 2: Sampling design III follows from sampling design II by conditioning on $X_{1.} = n_1,\ldots,X_{m.} = n_m$, and H_o^{II} transforms into H_o^{III}.

If we feel comfortable about conditional tests, we can treat all three designs as the same model. We shall also see that all tests and other statistical considerations run parallel for the three situations.

3. TWO-WAY CONTINGENCY TABLES

Since all three sample designs of the preceding section are identical except for conditioning, we shall only give a detailed

---

*The proofs of all Theorems have been omitted from this paper for the sake of brevity.

description of one of them. We start, therefore, with sample design I.

The model has likelihood function

$$(4) \quad L = \prod_i \prod_j (x_{ij}!)^{-1} \prod_i \prod_j \Pi\lambda_{ij}^{x_{ij}} \exp(-\sum_i \sum_j \lambda_{ij}).$$

We can write this on exponential form as

$$L = c^{-1}(\lambda_{11},\dots,\lambda_{mk}) \exp(\sum_i \sum_j x_{ij} \ln\lambda_{ij}) h(x_{11},\dots,x_{mk}),$$

with $c(\lambda_{11},\dots,\lambda_{mk}) = \exp\{\lambda_{..}\}$ and $h(x_{11},\dots,x_{mk}) = (\prod_i \prod_j x_{ij}!)^{-1}.$

The Poisson contengency table is thus an exponential family with canonical parameters $\theta_{11} = \ln\lambda_{11},\dots,\theta_{mk} = \ln\lambda_{mk}$, and sufficient statistics $x_{11},\dots,x_{mk}$.

The ML-estimators are given by the equations

$$x_{ij} = E[X_{ij}] = \lambda_{ij}, \quad i=1,\dots,m, j=1,\dots,k$$

with trivial solutions

$$(5) \quad \hat{\lambda}_{ij} = x_{ij}.$$

Under the hypothesis H_o^I of multiplicative mean values, we have

$$\ln L = C_1 \ (\lambda_{1.},\dots,\lambda_{m.},\lambda_{.1},\dots,\lambda_{.k}) +$$

$$\sum_{i=1}^{m-1} x_{i.} (\ln\lambda_{i.} -\ln\lambda_{m.}) \quad + \quad \sum_{j=1}^{k-1} x_{.j}(\ln\lambda_{.j}-\ln\lambda_{.k}) +$$

$$x_{..} (\ln\lambda_{..} +\ln\lambda_{m.} +\ln\lambda_{.k}) + \ln h(x_{11},\dots,x_{mk}),$$

where $C_1(\lambda_{1.},\dots,\lambda_{m.},\lambda_{.1},\dots,_{.k}) = -\lambda_{..}.$

Hence, we again have an exponential family with k-1+m-1+1 canonical parameters. The sufficient statistics are

$x_{i.}, i=1,\dots,m-1, x_{.j}, j=1,\dots,k-1$ and $x_{...}$ Hence, the ML-estimators follow from

$$x_{i.} = E[X_{i.}] = _{i.}, \quad i=1,\dots, m-1,$$

$$x_{.j} = E[X_{.j}] = \lambda_{.j}, \quad j=1,\ldots,k-1$$

and

$$x_{..} = E[X_{..}] = \lambda_{..},$$

with solutions

(6) $\hat{\lambda}_{i.} = x_{i.},$

(7) $\hat{\lambda}_{.j} = x_{.j}$

and

(8) $\hat{\lambda}_{..} = x_{..}$.

If we substitute (5) for λ_{ij} without H_o^I and (6) - (8) for $\lambda_{ij} = \lambda_{i.}\lambda_{.j}/\lambda_{..}$ under H_o, we get the LR-test statistic

(9) $z = -2\{\ln L(\hat{\lambda}_{1.}\hat{\lambda}_{.1}/\hat{\lambda}_{..},\ldots,\hat{\lambda}_{m.}\hat{\lambda}_{.k}/\hat{\lambda}_{..})$

$\qquad -\ln L(\hat{\lambda}_{11},\ldots,\hat{\lambda}_{mk})\}$

$= 2\Sigma_i \Sigma_j x_{ij}[\ln x_{ij} - \ln \dfrac{x_{i.}\,x_{.j}}{x_{..}}].$

We note that (9) has the familiar form $2\Sigma x_{ij}[\ln x_{ij} - \ln e_{ij}]$, where e_{ij} are the expected numbers under H_o^I. By theorem 3.6, z is asymptotically χ^2-distributed with r degrees of freedom, where r is the number of degrees of freedom specified under H_o. We have to be careful when we determine r. Under H_o^I, there are $m-1$ $\lambda_{i.}$'s, $k-1$ $\lambda_{.j}$'s and one $\lambda_{..}$, since given the value of $\lambda_{..}$, one $\lambda_{i.}$ and one $\lambda_{.j}$ is a function of the others. In the original model there are km parameters and we get

$$r = km-((k-1) + (m-1) +1) = (m-1)(k-1).$$

We reject H_o^I when the LR-test statistic is large. Hence the test with critical region

(10) $z > \chi^2_{1-\alpha}((m-1)(k-1))$

will have approximate level α. Since the test statistic only involves the observed numbers of the contingency table, we must expect it to have the same form for the other two designs. And this is in fact the case. As a help in understanding the principles of contingency tables, we show the necessary steps for design II, which is the most common.

The likelihood function is

(11) $L = (x_{1j}\ldots x_{mk})\prod_i \prod_j p_{ij}^{x_{ij}},$

or on exponential form

$$L = c^{-1}(p_{11}, \ldots, p_{mk}) \exp\{\sum_i \sum_j (\ln p_{ij} - \ln p_{mk}) x_{ij}\} h(x_{11}, \ldots, x_{mk}),$$

with $c(p_{11}, \ldots, p_{mk}) = p_{mk}^{-n}$ and $h(x_{11}, \ldots, x_{mk}) = (x_{11} \cdots x_{mk})^n$.

Hence the $mk-1$ canonical parameters are $\theta_{ij} = \ln p_{ij} - \ln p_{mk}$ for $(i,j) \neq (m,k)$ and the corresponding sufficient statistics are the x_{ij}'s. It follows that the ML-estimators are given by

$$x_{ij} = E[X_{ij}] = np_{ij},$$

or

(12) $\hat{p}_{ij} = x_{ij}/n.$

Under H_o^{II}, we get the likelihood function

$$L = c_1^{-1}(p_{1.}, \ldots, p_{m.}, p_{.1}, \ldots, p_{.k}) \exp\{\sum_{i=1}^{m-1}(\ln p_{i.} - \ln p_{m.}) x_{i.}$$

$$+ \sum_{j-1}^{k-1}(\ln p_{.j} - \ln p_{.k}) x_{.j}\} h(x_{11}, \ldots, x_{mk}),$$

with $c_1(p_{1.}, \ldots, p_{m.}, p_{.1}, \ldots, p_{.k}) = (p_{m.} p_{.k})^{-n}.$

Hence, the ML-estimators under H_o^I are given by

$$x_{i.} = E[X_{i.}] = np_{i.}$$

and

$$x_{.j} = E[X_{.j}] = np_{.j},$$

or

$$\hat{p}_{i.} = x_{i.}/n$$

and

$$\hat{p}_{.j} = x_{.j}/n.$$

With these ML-estimates under H_o^{II} and (12) without H_o^{II}, we then get the LR-test statistic

$$z = -2\{\ln L(\hat{p}_{1.}, \ldots, \hat{p}_{m.}, \hat{p}_{.1}, \ldots, \hat{p}_{.k}) - \ln L(\hat{p}_{11}, \ldots, \hat{p}_{mk})\}$$

$$= 2\sum_i \sum_j x_{ij}[\ln(x_{ij}/n) - \ln(\frac{x_{i.}}{n} \quad \frac{x_{.j}}{n})]$$

$$= 2\Sigma \sum_{i\ j} x_{ij}[\ln x_{ij} - \ln \frac{x_{i.} \, x_{.j}}{n}],$$

which with $n = x_{..}$ is identical with (9), and is 6 asymptotically χ^2-distributed with $r = (m-1)(k-1)$ degrees of freedom. The count of degrees of freedom is based on $mk-1$ canonical parameters in the original model and $m-1 + (k-1)$ canonical parameters under H_o^{II}. Hence, $mk-1-(m-1) - (k-1) = (m-1)(k-1)$ parameters are fixed under H_o^{II}. The critical region for the LR-test of H_o^{II} is accordingly given by (10) as for the LR-test for H_o^I.

We may summarize our findings in the following theorem:

Theorem 3: <u>The LR-test statistic for both H_o^I, H_o^{II} and H_o^{III} is given by</u>

$$z = 2 \sum_{i=1}^{m} \sum_{j=1}^{k} x_{ij}[\ln x_{ij} - \ln \frac{x_{i.} x_{.j}}{x_{..}}].$$

<u>A test of approximate level α has critical region</u>

$$z > \chi^2_{.95} \, ((k-1)(m-1)).$$

Example 1: As an application of the multiplicative poisson model and sample design I, we consider the data in Table 1. In 1962 a major investigation was carried out in Sweden to evaluate the influence of speed limits. As part of this investigation the number of accidents on Swedish roads were counted in 15 consecutive weeks. In addition, the roads were divided into groups: State highways, County roads and Other roads.

Table 1. Accidents On Swedish Roads Classified According To Type Of Road For 15 Weeks In 1962

Week	State Highways	County Roads	Other Roads	Total
1	2	7	4	13
2	8	8	4	20
3	7	9	9	25
4	7	4	8	19
5	3	5	7	15
6	5	4	4	13
7	4	5	7	16
8	4	4	12	20
9	7	3	8	18
10	3	8	12	23
11	4	12	15	31
12	4	5	14	23
13	9	12	10	31

Table 1.   (continued)

Week	State Highways	County Roads	Other Roads	Total
14	10	9	17	36
15	10	9	14	33
Total	87	104	145	336

We are going to test the multiplicative hypothesis

$$H_o^I : \lambda_{ij} = \lambda_{i.} \lambda_{.j}/\lambda_{..} ,$$

where λ_{ij} is the expected number of accidents in week i on road type j. If this hypothesis is accepted, we can describe the data by row effects $\lambda_{i.}$ and column effects $\lambda_{.j}$. The test for H_o^I gives

$$z = 27.3 , \quad df = 28.$$

Since $\chi^2_{.95}(28) = 41.3$, we can accept the hypothesis. In Table 2, the differences between observed and expected numbers

$$e_{ij} - x_{i.} x_{.j}/x_{..}$$

are shown.

We note, that the sign pattern does not show any systematic features, which confirms that x_{ij} is equal to $x_{i.} x_{.j}/x_{..}$ plus random errors.

Table 2.   Observed Minus Expected Numbers
For Swedish Accident Data

Week	State Highways	Country Roads	Other Roads
1	-1.4	+3.0	-1.6
2	+2.8	+1.8	-4.6
3	+0.5	+1.3	-1.8
4	+2.1	-1.9	-0.2
5	-0.9	+0.4	+0.5
6	+1.6	0.0	-1.6
7	-0.1	0.0	-1.0
8	-1.2	-2.2	+3.4
9	+2.3	-2.6	+0.2
10	-3.0	+0.8	+2.1
11	-4.0	+2.4	+1.6
12	-2.0	-2.1	+4.1
13	+1.0	+2.4	-3.4

Table 2. (continued)

Week	State Highways	Country Roads	Other Roads
14	+0.7	-2.1	+1.5
15	+1.5	-1.2	-0.2

Now that H_o^I is accepted, the parameters of the model are λ_1, $\ldots,\lambda_m$, $\lambda_1,\ldots,\lambda_k$ and λ. The estimates are given by (6) (7) and (8), or simply the marginals of Table 1. These estimates show that independently of road type, the number of accidents has risen during the 15 weeks. The column totals show that indepently of the week, there is an increasing number as we go to smaller roads, seem- ing to indicate that small roads are more dangerous than highways. But this is, of course, not true, since we have not taken into account that there are likely to be many more miles of small roads. This shows that great care must be exercised in interpreting par- ameters.

4. LOG-LINEAR MODELS

The treatment of contingency tables becomes much more elegant, if we make use of a new parametrization termed log-linear model.

Whether we have sample design I, II or III, we write the model as

$$\mu_{ij}= E[X_{ij}] = \exp\{\theta_{ij}^{(12)} + \theta_i^{(1)} + \theta_j^{(2)} + \theta^{(o)}\}$$

or

(13) $\mu_{ij}^{*}= \ln \mu_{ij} = \theta_{ij}^{(12)}+ \theta_i^{(1)} + \theta_j^{(2)} + \theta^{(o)}.$

The parameters must satisfy the constraints

(14) $\theta_{i.}^{(12)} = \theta_{.j}^{(12)} = \theta_.^{(1)} = \theta_.^{(2)} = 0,$

i.e. all log-linear model parameters must sum to zero over all indices. We note that the model has exactly mk parameters under the constraints (14), since we have $(k-1)(m-1)$ $\theta_{ij}^{(12)}$'s, $(m-1)$ $\theta_i^{(1)}$'s, $(k-1)$ $\theta_j^{(2)}$'s and one $\theta^{(o)}$.

That (13) is in fact a reparametrization follows from the easily verified formulas

$$\theta_{ij}^{(12)} = \mu_{ij}^{*} -\mu_{i.}^{*} -\mu_{.j}^{*} + \mu_{..}^{*},$$

$$\theta_i^{(1)} = \bar{\mu}_{i.}^* - \bar{\mu}_{..}^* ,$$

$$\theta_j^{(2)} = \bar{\mu}_{.j}^* - \bar{\mu}_{..}^*$$

and

$$\theta^{(o)} = \bar{\mu}_{..}^* ,$$

where $\bar{\mu}_{i.}^* = \dfrac{1}{k} \underset{j}{\Sigma} \mu_{ij}^*,$ $\bar{\mu}_{.j}^* = \dfrac{1}{m} \underset{i}{\Sigma} \mu_{ij}^*$ and $\bar{\mu}_{..}^* = \dfrac{1}{km} \underset{ij}{\Sigma\Sigma} \mu_{ij}^*.$

We now have the following important result

<u>Theorem 4:</u> <u>The hypothesis H$_{12}$:</u> $\theta_{ij}^{(12)} = 0$ <u>for all i</u> <u>and j is</u> <u>identical with</u> H_o^I, H_o^{II} or H_o^{III} .

We have special names for the parameters of a log-linear model. $\theta_{ij}^{(12)}$ are called <u>interactions</u>, $\theta_i^{(1)}$ are called <u>row effects</u>, $\theta_j^{(2)}$ are called <u>column effects</u> and $\theta^{(o)}$ is called the <u>over all effect</u>.

From (4) it follows that the log-likelihood function under Poisson sampling has the form

(15) $\ln L = \underset{i}{\Sigma} \underset{j}{\Sigma} x_{ij} \ln \lambda_{ij} - \lambda_{..} - h_1(x_{11}, \dots, x_{mk})$

$$= \underset{ij}{\Sigma\Sigma} x_{ij} \{ \theta_{ij}^{(12)} + \theta_i^{(1)} + \theta_j^{(2)} + \theta^{(o)} \}$$
$$- \Sigma\Sigma \exp\{ \theta_{ij}^{(12)} + \theta_i^{(1)} + \theta_j^{(2)} + \theta^{(o)} \}$$

$$- h_1(x_{11}, \dots, x_{mk}) ,$$

where $h_1(x_{11}, \dots, x_{mk}) = \underset{i}{\Sigma} \underset{j}{\Sigma} \ln x_{ij}!.$

Under sample design II, the log-likelihood function follows from (11) and we get

(16) $\ln L = \underset{i}{\Sigma} \underset{j}{\Sigma} x_{ij} \ln(np_{ij}) - h_2(x_{11}, \dots, x_{mk})$
$$= \underset{i}{\Sigma} \underset{j}{\Sigma} x_{ij} \{ \theta_{ij}^{(12)} + \theta_1^{(1)} + \theta_j^{(2)} + \theta^{(o)} \} - h_2(x_{11}, \dots, x_{mk}) ,$$

where $h_2(x_{11}, \dots, x_{mk}) = -\ln(x_{11} \overset{n}{\dots} x_{mk}) + n \ln n.$

Under sample design III, the likelihood function is

$$L = \prod_{i=1}^{m} \{ (x_{i1} \overset{n_i}{\cdots} x_{ik}) \, (p^*_{i1})^{x_{i1}} \cdots (p^*_{ik})^{x_{ik}} \}$$

where p^*_{ij} is defined as in Section 2.

Hence

$$(17) \quad \ln L = \sum_{i=1}^{m} \sum_{j=1}^{k} x_{ij} \ln(n_i p^*_{ij}) - h_3(x_{11}, \ldots, x_{mk})$$

$$= \sum_i \sum_j x_{ij} \{ \theta^{(12)}_{ij} + \theta^{(1)}_i + \theta^{(2)}_j + \theta^{(0)} \} - h_3(x_{11}, \ldots, x_{mk}),$$

where $h_3(x_{11}, \ldots, x_{mk}) = -\sum_i \ln\left(x_{i1} \overset{n_i}{\cdots} x_{ij}\right) + \sum_i n_i \ln n_i$.

Thus in all three cases, we have the exponential form

$$(18) \quad \ln L = \sum_i \sum_j x_{ij} \theta^{(12)}_{ij} + \sum_i x_{i.} \theta^{(1)}_i$$

$$+ \sum_j x_{.j} \theta^{(2)}_j + x_{..} \theta^{(0)} - c(\theta) - h(\underline{x}),$$

where $c(\theta)$ is a function of the various θ's and $h(\underline{x})$ a function of the x_{ij}'s.

Equation (18) shows that the model under all three sampling designs forms an exponential family with canonical parameters $\theta^{(12)}_{ij}$, $\theta^{(1)}_i$, $\theta^{(2)}_j$ and $\theta^{(0)}$. The corresponding minimal sufficient statistics are x_{ij}, $x_{i.}$, $x_{.j}$ and $x_{..}$. Whichever design we now have, the fundamental set of likelihood equations is

$$(19) \quad x_{ij} = E[X_{ij}],$$

$$(20) \quad x_{i.} = E[X_{i.}],$$

$$(21) \quad x_{.j} = E[X_{.j}]$$

and

$$(22) \quad x_{..} = E[X_{..}].$$

Since there are only $(k-1)(m-1)$ canonical parameters leading to (19), we only have to solve (19) for $i=1,\ldots,m-1, j=1,\ldots k-1$. Correspondingly we only have to solve (20) for $i=1,\ldots,m-1$ and (21) for $j=1,\ldots,k-1$. It is easy to see, on the other hand, that (20), (21) and (22) are automatically satisfied if (19) is satisfied for all i and j. Hence, we can simply solve (19) for all i and j,

giving us the complete solution for all parameters. This will often be the case with log-linear models in later sections. Several of the equations will automatically be satisfied if others are satisfied. Any set of equations that give us a complete solution is called a sufficient set of equations. Since the ML-equations are always of the form: "some marginals equal to their mean-values," we term the marginals of a sufficient set of equations: <u>sufficient marginals</u>. For the log-linear model (13), we thus find that the x_{ij}'s are the sufficient marginals.
(The observations x_{ij} are not really marginals, but it will become clear in the following that it is natural to think of them as marginals.

The solutions to (19) follow from the following theorem.

<u>Theorem 5</u>:   <u>The ML-estimates for the log-linear model are given by</u>

(23) $\hat{\theta}_{ij}^{(12)} = x_{ij}^* - \bar{x}_{i.}^* - \bar{x}_{.j}^* + \bar{x}_{..}^*$,

(24) $\hat{\theta}_i^{(1)} = \bar{x}_{i.}^* - \bar{x}_{..}^*$,

(25) $\hat{\theta}_j^{(2)} = \bar{x}_{.j}^* - \bar{x}_{..}^*$,

and

(26) $\hat{\theta}_o = \bar{x}_{..}^*$,

<u>where</u> $x_{ij}^* = \ln x_{ij}$.   <u>The solutions exist provided</u> $x_{ij} \neq 0$ <u>for all i and j.</u>

From theorem 5, we have explicit estimates for the interaction parameters of a two-way contingency table. If the hypothesis of no interactions is rejected, the natural question is which of the interactions are actually different from 0. To help answer this question, we have

<u>Theorem 6</u>:   <u>The interaction parameter estimates</u>

$\hat{\theta}_{ij}^{(12)}$, i=1,...,m-1, j=1,..., k-1

<u>are asymptotically normally distributed with large sample variances</u>

(27) $\text{var}\left[\hat{\theta}_{ij}^{(12)}\right] \simeq \dfrac{(k-2)(m-2)}{mk} \dfrac{1}{np_{ij}}$

$+ \dfrac{1}{m^2} \dfrac{k-2}{k} \sum_{r=1}^{m} \dfrac{1}{np_{ij}}$

$$+ \frac{1}{k^2} \frac{m-2}{m} \sum_{t=1}^{k} \frac{1}{np_{it}} + (\frac{1}{mk})^2 \sum_{r=1}^{m} \sum_{t=1}^{k} \frac{1}{np_{rt}}$$

It follows from theorem 6 that standard errors of the inter-action estimates are easily computed. This means that we can judge the influence of the $\theta_{ij}^{(12)}$'s by the quantity

$$U_{ij} = \hat{\theta}_{ij}^{(12)} / \sqrt{\text{var}[\hat{\theta}_{ij}^{(12)}]} \ ,$$

which is asymptotically distributed as $N(0,1)$. An estimate for U_{ij} would be

(28) $\hat{u}_{ij} = \hat{\theta}_{ij}^{(12)} / \hat{\sigma}_{ij}^{(12)}$

where $(\hat{\sigma}_{ij}^{(12)})^2$ is var $[\hat{\theta}_{ij}^{(12)}]$ with np_{ij} replaced by x_{ij}.

If the hypothesis

(29) H_{12}: $\theta_{ij}^{(12)} = 0$ for all i and j

has been rejected, we may thus construct a table of the $\hat{u}_{ij}$'s to get an idea of which cells are likely to have $\theta_{ij} = 0$ and which cells are likely to have $\theta_{ij} \neq 0$. To do so, we simply note which cells have a value of $\hat{u}_{ij}$ larger than 1.96.

Assume now that H_{12} given by (29) has been accepted. We then want to estimate the remaining parameters $\theta_i^{(1)}$, $\theta_j^{(2)}$ and $\theta^{(o)}$. From the exponential form (15), (16) or (17) for the three designs, we note that when $\theta_{ij}^{(12)} = 0$ for all i and j, and the likelihood equations are simply (20), (21) and (22). But since (22) follows from (20) for all i as well as from (21) for all j, (20), and (21) form a set of sufficient equations, and the vectors $(x_{1.},...,x_{m.})$ and $(x_{.1},...,x_{.k})$ accordingly are the sufficient marginals. Hence we easily get

Theorem 7: Under H_{12}: $\theta_{ij}^{(12)} = 0$ for all i and j, the ML-estimates are

(30) $\hat{\theta}_i^{(1)} = \ln x_{i.} - m^{-1} \sum_i \ln x_{i.}$

(31) $\hat{\theta}_j^{(2)} = \ln x_{.j} - k^{-1} \sum_j \ln x_{.j}$

and
(32) $\hat{\theta}^{(o)} = m^{-1} \sum_i \ln x_{i.} + k^{-1} \sum_j \ln x_{.j} - \ln x$

Note that the estimates of the row and column effects are different under H_{12} and without H_{12}.

The reader may find that little new knowledge has been gained in this section, and he may be partly correct. As we shall see in the following sections, however, the log-linear model approach makes the generalizations to contingency tables of higher dimensions almost trivial. This is not the case with the traditional parametrization.

Before we proceed to 3 and 4 dimensions, we shall, however, discuss tests concerning row and column effects.

Sometimes we have to consider contingency tables with incomplete data. Examples of such cases are tables, where some cells have zero counts and tables, where two or more cells are mixed up. We can thus have two-dimensional tables, where $x_{ij} = o$ for some combinations of i and j, and we can have tables, where we only know certain partial sums as, for example,

$$x_{i1} + \cdots + x_{ir} = x_i^{(1)}$$

and

$$x_{ir+1} + \cdots + x_{ik} = x_i^{(2)}.$$

Sometimes the parameters of the model can still be estimated; sometimes there is not enough information in the data to estimate certain parameters.

5. TESTS FOR ROW AND COLUMN EFFECTS

If the z-test has shown that all the interactions are 0, we may be interested in testing hypotheses concerning the row and column effects. The most common tests are

(33) $H_1: \quad \theta_i^{(1)} = 0$ for all i

and

(34) $H_2: \quad \theta_j^{(2)} = 0$ for all j.

If H_1 holds, we say that there are no row effects and if H_2 holds we say that there are no column effects.

It is interesting to note what H_1 and H_2 correspond to for sample designs I, II and III.

a) For sample design I, we have the alternative forms of H_1 and H_2:

(35) H_1: $\lambda_{i.} = \lambda_{..}/m$ for all i

and
(36) H_2: $\lambda_{.j} = \lambda_{..}/k$ for all j.

These results are easily verified since under H_{12} we have

$$\lambda_{ij} = E[X_{ij}] = \exp\{\theta_i^{(1)} + \theta_j^{(2)} + \theta^{(o)}\}$$

Under H_1 we then have
$$\lambda_{i.} = \exp\{\theta^{(o)}\}\Sigma_j \exp\{\theta_j^{(2)}\} \ ,$$

and

$$\lambda_{..} = m \exp\{\theta^{(o)}\}\Sigma_j \exp\{\theta_j^{(2)}\},$$

from which (35) follows. We can establish (36) in the same way.
b) For sample design II, we have

(37) H_1: $P_{i.} = \dfrac{1}{m}$, $i = 1,\ldots,m,$

and

(38) H_2: $P_{.j} = \dfrac{1}{k}$, $j = 1,\ldots,k.$

Since under H_{12}
$$np_{ij} = \exp\{\theta_i^{(1)} + \theta_j^{(2)} + \theta^{(o)}\},$$

we get under H_1
$$np_{i.} = \exp\{\theta^{(o)}\}\Sigma_j \exp\{\theta_j^{(2)}\}$$

and since $\Sigma_i p_{i.} = 1$

$$n = \exp\{\theta^{(o)}\}\Sigma_j \exp\{\theta_j^{(2)}\}.m$$

Hence,
$$P_{i.} = \frac{1}{m,}\qquad\qquad i = 1,\ldots,m,$$

which proves (37). (38) is shown in the same manner.

The hypothesis of no row effects thus corresponds to the hypothesis that all marginal row probabilities are equal.

c) For sample design III, the row marginals are fixed and hence only H_2 makes sense. We easily prove as under a) and b) that H_2 is equivalent to

(39) $H_2 : p_{.j}^* = \dfrac{1}{k}$ for all j.

Under H_1, only (21) and (22) remain as likelihood equations. And since (22) is satisfied when (21) is satisfied for all j, it follows that the vector $(x_{.1}, \ldots, x_{.k})$ forms the set of sufficient marginals under H_1. Under H_2, we have, in the same way that $(x_{1.}, \ldots, x_{m.})$ are the sufficient marginals. The actual estimates follow from

Theorem 8: Under H_1 the ML-estimates are

$$\hat{\theta}_j^{(2)} = \ln x_{.j} - \frac{1}{k} \sum_j \ln x_{.j}$$

and

$$\hat{\theta}^{(o)} = \frac{1}{k} \sum_j \ln x_{.j} - \ln(m).$$

Under H_2, the ML-estimates are

$$\hat{\theta}_i^{(1)} = \ln x_{i.} - \frac{1}{m} \sum_i \ln x_{i.}$$

and

$$\theta^{(o)} = \frac{1}{m} \sum_i \ln x_{i.} - \ln(k).$$

The tests for H_1 and H_2 are easily set up. We have (under H_{12})

Theorem 9: The LR-test statistic for H_1 against H_{12} is

(40) $Z_1 = 2 \sum\limits_{i=1}^{m} X_{i.} \{\ln X_{i.} - \ln(\dfrac{X_{..}}{m})\}$,

which is asymptotically $\chi^2(m-1)$ under H_1 , and the LR-test statistic for H_2 against H_{12} is

(41) $Z_2 = \sum\limits_{j=1}^{k} X_{.j} \{\ln X_{.j} - \ln(\dfrac{X_{..}}{k})\}$,

which is asymptotically $\chi^2(k-1)$ under H_2.

The tests described in theorems 3 and 9, can be summarized in a variation table, which is very similar to the tables of variation applied for analysis of variance methods. The table takes the form shown on Table 3.

Table 3. Variation Table For The Analysis Of A Two-Way Contingency
Table

Variation	Hypothesis	Test Statistic	Degrees of Freedom
Between cells	H_{12}	Z	$(m-1)(k-1)$
Between rows	H_1	Z_1	$m-1$
Between columns	H_2	Z_2	$k-1$
Total		Z_o	$mk-1$

The test statistic

$$Z_o = \underset{ij}{\Sigma\Sigma} X_{ij}\{\ln X_{ij} - \ln \frac{X_{..}}{mk}\}$$

with df = mk-1 is interesting. Z_o makes a direct test of all three
hypotheses H_{12}, H_1, H_2 at the same time, since the expected numbers
under H_{12}, H_1 and H_2 are

$$X_{..}/(mk).$$

Before considering an example, we make a few remarks. Note firstly
that Z_1 is the test for equal cell probabilities in the multinomial
distribution of $(X_1,...,X_m)$. Hence we could have obtained the
test statistic directly if we had been sure that we could restrict
attention to the marginal distribution of $(X_1,...,X_m)$ under H_{12}.
This is an example of the <u>sufficientcy principle</u>. If a sufficient
statistic exist, then attention can be restricted to the sufficient
statistic and we must have that the LR-test statistic is a function
of it. But under H_{12}, X_i is minimal sufficient for $\theta_i^{(1)}$ and the
sufficiency principle thus applies.

We also realize that in Table 3 we carry out a sequence of
tests. Thus H_1 and H_2 are tested under the assumption that H_{12}
has been accepted. This is a very important philosophy of contingency table methods.

We order all the hypotheses, we intend to test, in a so-called
<u>hierarchical order</u>, which tells us which hypotheses to test first
and which hypotheses are tested under the assumption that other
hypotheses hold. For Table 3, the hierarchical order is

$$H_{12} \begin{matrix} \nearrow H_1 \\ \searrow H_2 \end{matrix}$$

Thus H_1 and H_2 are parallel in the ordering.

When we come to the three-dimensional tables, we shall pursue the subject of hierarchical order in much more detail.

Example 2: Let us apply the developed technique to the traffic data of Table 1. We have already accepted H_{12}. But we want to analyze the row effects and column effects. In Table 4 the estimates of column and row effects under H_{12}, given by (30) and (31) are shown.

Table 4. Estimates Of Row And Column Effects For Traffic Data

	i= 1	2	3				
$\hat{\theta}_i^{(1)} =$	-0.23	-0.06	+0.28				

	j= 1	2	3	4	5	6	7
$\hat{\theta}_j^{(2)} =$	-0.50	-0.06	+0.16	-0.12	-0.35	-0.50	-0.29

	j= 8	9	10	11	12	13	14	15
$\hat{\theta}_j^{(2)} =$	-0.16	-0.17	+0.08	+0.37	+0.08	+0.37	+0.52	+0.44

For both row and column effects, there seems to be a systematic increase with i and j. The table of variation shown in Table 5 confirms this and shows that the differences cannot be considered as random errors.

Table 5. Table Of Variation For Traffic Data

Variation	Hypothesis	z	df
Between cells	$\theta_{ij}^{(12)} = 0$	27.3	28
Between rows	$\theta_i^{(1)} = 0$	33.7	14
Between columns	$\theta_j^{(2)} = 0$	15.5	2
Total	all $= 0$	76.5	44

As we can see both H_1 and H_2 are rejected. The rejection of H_1 shows that there has been a significant increase in the number of accidents independently of road type. The rejection of H_2 does not tell us much. We do not know why or even if there are significantly more accidents on small roads. We must first get information on the total length of the various roads.

6.   THREE-WAY CONTINGENCY TABLES

We shall deal with three-way contingency tables in a systematic way such that the generalizations to higher dimensions become obvious. In doing so, we shall work almost exclusively with the log-linear form and only refer to the specific models occasionally and then only for sample design II, which is the most commonly used. Note that in three dimensions there are more than three possible designs since several marginals may be predetermined.

Consider then a contingency table x_{ijl}, $i = 1,\ldots,m$, $j = 1,\ldots k$, $l = 1,\ldots,r$, where the mean values are given in log-linear form

$$(42) \quad \ln E[X_{ijl}] = \theta_{ijl}^{(123)} + \theta_{ij}^{(12)} + \theta_{il}^{(13)} + \theta_{jl}^{(23)}$$
$$+ \theta_i^{(1)} + \theta_j^{(2)} + \theta_l^{(3)} + \theta^{(o)} \ .$$

The following constraints are imposed

$$(43) \quad \theta_{ij.}^{(123)} = \theta_{i.l}^{(123)} = \theta_{.jl}^{(123)} = 0 \ ,$$

$$(44) \quad \theta_{i.}^{(12)} = \theta_{.j}^{(12)} = \theta_{i.}^{(13)} = \theta_{.l}^{(13)} = \theta_{j.}^{(23)} = \theta_{.l}^{(23)} = 0$$

and

$$(45) \quad \theta_.^{(1)} = \theta_.^{(2)} = \theta_.^{(3)} = 0 \ .$$

The fact that (42) represents a proper reparametrization can be seen in two steps. Firstly, we simply count the number of θ's and compare with the mkr parameters of the original model. Due to the constraints (43), (44) and (45), we only have the number of free parameters shown as column 3 in Table 6. The table also shows the names attached to the various parameters. Note that we use the common name, main effect, for what corresponds to row effects and column effects in two-way tables.

Secondly, we may explicitly write down the θ parameters in terms of the $\log E[X_{ijl}]$'s, which under all designs are direct functions of the parameters. It is not difficult to verify that, with $\mu_{ijl} = \ln E[X_{ijl}]$, we have

Table 6: Summary Of The Log-Linear Model Parameters

Parameter	Name	Number
$\theta_{ijl}^{(123)}$	3rd order interaction	$(m-1)(k-1)(r-1)$
$\theta_{ij}^{(12)}$	2nd order interaction	$(m-1)(k-1)$
$\theta_{il}^{(13)}$	2nd order interaction	$(m-1)(r-1)$
$\theta_{jl}^{(23)}$	2nd order interaction	$(k-1)(r-1)$
$\theta_{i}^{(1)}$	main effect	$m-1$
$\theta_{j}^{(2)}$	main effect	$k-1$
$\theta_{l}^{(3)}$	main effect	$r-1$
$\theta^{(o)}$	over all level	1

Total		mkr

$$\theta_{ijl}^{(123)} = \bar{\mu}_{ijl}^{*} - \bar{\mu}_{ij.}^{*} - \bar{\mu}_{i.l}^{*} - \bar{\mu}_{.jl}^{*} + \bar{\mu}_{i..}^{*} + \bar{\mu}_{.j.}^{*} + \bar{\mu}_{..l}^{*} - \bar{\mu}_{...}^{*} ,$$

$$\theta_{ij}^{(12)} = \bar{\mu}_{ij.}^{*} - \bar{\mu}_{i..}^{*} - \bar{\mu}_{.j.}^{*} + \bar{\mu}_{...}^{*} ,$$

$$\theta_{i}^{(1)} = \bar{\mu}_{i..}^{*} - \bar{\mu}_{...}^{*} ,$$

and

$$\theta^{(o)} = \bar{\mu}_{...}^{*} ,$$

where a bar means an average. The expressions for

$\theta_{il}^{(13)}$, $\theta_{jl}^{(23)}$, $\theta_{j}^{(2)}$ and $\theta_{l}^{(3)}$ are similar.

We cannot, from (42) see directly that we have an exponential family. To satisfy ourselves that this is in fact the case, we write down the log-likelihood function for sample designs I and II. For design I, the x_{ijl}'s are independent Poisson distributed. Accordingly we have

$$(46) \quad \ln L = \underset{ijl}{\Sigma\Sigma\Sigma} x_{ijl} \ln \lambda_{ijl} - \underset{ijl}{\Sigma\Sigma\Sigma} \lambda_{ijl} - \underset{ijl}{\Sigma\Sigma\Sigma} \ln \left(x_{ijl}! \right)$$

$$= \Sigma\Sigma\Sigma x_{ijl}\{\theta_{ijl}^{(123)} + \theta_{ij}^{(12)} + \theta_{il}^{(13)}$$

$$+\theta_{jl}^{(23)} + \theta_i^{(1)} + \theta_j^{(2)} + \theta_l^{(3)} + \theta^{(o)}\}$$

$$-\lambda_{...} - \Sigma\Sigma\Sigma \ln\left(x_{ijl!}\right),$$

such that the θ_{ijl}'s are canonical parameters of our exponential family. For sample design II, we have

$$(47) \quad \ln L = \Sigma\Sigma\Sigma x_{ijl}\ln p_{ijl} + \ln\left(x_{111}\overset{n}{\cdots}x_{mkr}\right),$$

$$= \Sigma\Sigma\Sigma x_{ijl}\{\theta_{ijl}^{(123)} + \theta_{ij}^{(12)} + \theta_{il}^{(13)} + \theta_{jl}^{(23)}$$

$$+ \theta_i^{(1)} + \theta_j^{(2)} + \theta_l^{(3)} + \theta^{(o)}\} - \ln(n)$$

$$+ \ln\left(x_{111}\overset{n}{\cdots}x_{mkr}\right),$$

which apart from the terms dependent only on parameters or observations is identical with the poisson log-likelihood function. In order to identify the sufficient statistics and the likelihood equations, we only have to write out the part of lnL containing θ's ans x's together. From either (46) or (47), we get

$$\Sigma\Sigma\Sigma x_{ijl}\{\theta_{ijl}^{(123)} + \theta_{ij}^{(12)} + \theta_{il}^{(13)} + \theta_{jl}^{(23)}$$

$$+\theta_i^{(1)} + \theta_j^{(2)} + \theta_l^{(3)} + \theta^{(o)}\} = \Sigma\Sigma\Sigma x_{ijl}\theta_{ijl}^{(123)} + \Sigma\Sigma\theta_{ij}^{(12)}x_{ij.}$$

$$+\Sigma\Sigma\theta_{il}^{(13)}x_{i.l} + \Sigma\Sigma\theta_{jl}^{(23)}x_{.jl}$$

$$+\Sigma\theta_i^{(1)}x_{i..} + \Sigma\theta_j^{(2)}x_{.j.} + \Sigma\theta_l^{(3)}x_{..l} + \theta^{(o)}x_{...}.$$

It follows that the various marginals are sufficient for the corresponding canonical parameters. This leads to the likelihood equations

$$(48) \quad x_{ijl} = E[X_{ijl}], \quad [\theta_{ijl}^{(123)}],$$

(49) $x_{ij.} = E[X_{ij.}]$, $[\theta_{ij}^{(12)}]$,

(50) $x_{i.1} = E[X_{i.1}]$, $[\theta_{i1}^{(13)}]$,

(51) $x_{.j1} = E[X_{.j1}]$, $[\theta_{j1}^{(23)}]$,

(52) $x_{i..} = E[X_{i..}]$, $[\theta_{i}^{(1)}]$

(53) $x_{.j.} = E[X_{.j.}]$, $[\theta_{j}^{(2)}]$

(54) $x_{..1} = E[X_{..1}]$, $[\theta_{1}^{(3)}]$ and

(55) $x_{...} = E[X_{...}]$, $[\theta^{(o)}]$

In the brackets to the right are shown the cannonical parameters corresponding to the equations. If all parameters are to be estimated, x_{ijl} for all i, j and l forms the set of sufficient marginals, since (49) to (50) are consequences of (48).

We shall try to avoid further technical complications by omitting attempts to solve likelihood equations in the following. A number of computer programs are available which estimate all canonical parameters if the corresponding, sufficient marginals are specified. Accordingly, we shall, in the following specify the sufficient marginals under any hypothesis and then assume that the estimates are computed by means of a computer program.

7. HYPOTHESIS TESTING IN THREE-WAY CONTINGENCY TABLES

For contingency tables of high order, it is essential that we organize the hypotheses to be tested in a systematic fashion. We can formulate a large number of hypotheses based on the kmr canonical parameters of the model. Even if we restrict our attention to those hypotheses where one or more canonical parameters are 0, there are a substantial number of different hypotheses to be tested, namely 2^{mkr}. (For the most simple table with m=k=r=2, there are as many as 64 different hypotheses). In the following, we shall aim at two goals. Firstly, we shall limit ourselves to groups of hypotheses that have a meaningful interpretation. Secondly, we shall discuss the order in which it makes sense to carry out the hypothesis testing.

We start by listing the following set of basic hypotheses.

H_{123}: $\theta_{ijl}^{(123)} = 0$ for all i, j and l.

H_{12}: $\theta_{ijl}^{(123)} = 0$ and $\theta_{ij}^{(12)} = 0$ for all i, j and l.

H_{13}: $\theta_{ijl}^{(123)} = 0$ and $\theta_{il}^{(13)} = 0$ for all i, j and l.

H_{23}: $\theta_{ijl}^{(123)} = 0$ and $\theta_{jl}^{(23)} = 0$ for all i, j and l.

H_1: All interactions of 2nd and 3rd order are 0 and

$\theta_i^{(1)} = 0$ for all i.

H_2: All interactions of 2nd and 3rd order are 0 and

$\theta_j^{(2)} = 0$ for all j.

H_3: All interactions of 2nd and 3rd order are 0 and $\theta_l^{(3)} = 0$ for all l.

H_o: All parameters are 0 except θ .

These hypotheses can be combined in various ways to give new relevant hypotheses. The most important hypotheses are H_{12}, H_{13} and H_{23}. The interpretation of H_{12} follows from the following theorem.

Theorem 10: Hypothesis H_{12} is equivalent to

(56) $\quad P_{ijl} = P_{i.l}P_{.jl}/P_{..l}$

for design II, or

(57) $\quad \lambda_{ijl} = \lambda_{i.l}\lambda_{.jl}/\lambda_{..l}$

for design I.

Condition (56) can be interpreted as conditional independence of criteria 1 and 2 given the category under criterion 3. Let

A_i = {an individual belongs to category i of criterion 1} ,

B_j = {an individual belongs to category j of criterion 2} and

C_l = {an individual belongs to category l of criterion 3}

Then $P\{A_i|C_l\} = P_{i.l}/P_{..l}$, $P\{B_j|C_l\}= P_{.jl}/P_{..l}$ and $P\{A_i \cap B_j|C_l\}$

$= P_{ijl}/P_{..l}$

Conditional independence of criteria 1 and 2 given category 1 of criterion 3 is thus identical with

$$P_{ij1}/P_{..1} = (P_{i.1}/P_{..1})(P_{.j1}/P_{..1}), \quad \text{and this is precisely}$$

condition (56).

If H' -> H'' means that H' holds if H'' holds, we have the route diagram

Thus, we cannot proceed to H_1, H_2 or H_3 before either H_{12}, H_{13} or H_{23} has been accepted. It has been suggested to test the various hypotheses as successive tests, such that each new hypothesis assumes all previous ones hold. Such sequences of hypotheses are easily constructed if we introduce

$H_{13,23}$: both H_{13} and H_{23} hold, or $\theta_{ij1}^{(123)} = 0$, $\theta_{il}^{(13)} = 0$, $\theta_{j1}^{(23)} = 0$.

$H_{12,23}$: both H_{12} and H_{23} hold, or $\theta_{ij1}^{(123)} = 0$, $\theta_{ij}^{(12)} = 0$, $\theta_{j1}^{(23)} = 0$.

$H_{12,13}$: both H_{12} and H_{13} hold, or $\theta_{ij1}^{(123)} = 0$, $\theta_{ij}^{(12)} = 0$, $\theta_{il}^{(13)} = 0$.

$H_{12,13,23}$: both H_{12}, H_{13} and H_{23} holds, or all interactions of 2nd order and 3rd order are 0.

$H_{2,3}$: both H_2 and H_3 holds, or all parameters are 0, except $\theta_i^{(1)}$ and $\theta^{(o)}$.

$H_{1,3}$: both H_1 and H_3 holds, or all parameters are 0, except $\theta_j^{(2)}$ and $\theta^{(o)}$.

$H_{1,2}$: both H_1 and H_2 holds, or all parameters are 0, except $\theta_1^{(3)}$ and $\theta^{(o)}$.

$H_{1,2,3}$: both H_1, H_2 and H_3 holds, or all parameters are 0 except $\theta^{(o)}$.

The interpretation of $H_{12,13}$, $H_{13,23}$ follows from the following theorem ($H_{13,23}$ directly, the others by exchange of indices):

Theorem 11: If $H_{13,23}$ holds then

(58) $P_{i.1} = P_{i..} P_{..1}$ and (59) $P_{.j1} = P_{.j.} P_{..1}$

under design II, i.e. criteria 1 and 3, as well as criteria 2 and 3, are independent. For design I, $H_{13,23}$ implies that

$$\lambda_{i.1} = \lambda_{i..} \lambda_{..1}/\lambda_{...} \quad \text{and} \quad \lambda_{.j1} = \lambda_{.j.} \lambda_{..1}/\lambda_{...} .$$

According to theorem 11 criteria 1 and 3 are independent, and criteria 2 and 3 are independent if $H_{13,23}$ is true. In this situation the only interaction between criteria is between criteria 1 and 2.

We can now construct various successive systems of hypotheses. One example is

(60) $\quad H_{123} \longrightarrow H_{12} \longrightarrow H_{12,13} \longrightarrow H_{12,13,23}$

$\quad\quad \longrightarrow H_1 \longrightarrow H_{1,2} \longrightarrow H_{1,2,3} .$

In the following we consider only successive testing. The chain (60) means that we test hypotheses concerning the parameters is the following order, where each step assumes taht all previous hypotheses have been accepted:

a) $\quad \theta_{ij1}^{(123)} = 0,$

b) $\quad \theta_{ij}^{(12)} = 0,$

c) $\quad \theta_{i1}^{(13)} = 0,$

d) $\quad \theta_{j1}^{(23)} = 0,$

e) $\quad \theta_{i}^{(1)} = 0,$

f) $\quad \theta_{j}^{(2)} = 0 \quad$ and

g) $\quad \theta_{1}^{(3)} = 0.$

The chain (60) is accordingly based on the assumption that the 2nd order interactions between criteria 1 and 2 are most likely 0, next most likely to be 0 are the interactions between criteria 1 and 3 and so on. Unfortunately, there are no rules for how to select the sequence of successive hypotheses to be tested.

For use in the test quantities, we need the ML-estimates for the non-null parameters under each hypothesis. As we have seen, it is enough to specify the sufficient marginals which tell us which likelihood equations to solve. We also note that several of the equations follow from other equations. In Table 7 all the hypothesis, the sufficient marginals and the sufficient equations are listed. For convenience only the superscripts of the 0's are listed.

Table 7. Summary Of Hypotheses And Sufficient Marginals

Hypothesis	Number of parameters	Equations	Sufficient Marginals
H_{123}	$mkr-(m-1)(k-1)(r-1)$	52,53,54	(12),(13),(23)
H_{12}	$r(m+k-1)$	53,54	(13),(23)
H_{13}	$k(m+r-1)$	52,54	(12),(23)
H_{23}	$m(k+r-1)$	52,53	(12),(13)
$H_{13,23}$	$mk+r-1$	52,57	(12),(3)
$H_{12,23}$	$mr+k-1$	53,56	(13),(2)
$H_{13,12}$	$kr+m-1$	54,55	(23),(1)
$H_{12,13,23}$	$m+k+r-2$	55,56,57	(1),(2),(3)
H_1	$k+r-1$	56,57	(2),(3)
H_2	$m+r-1$	55,57	(1),(3)
H_3	$k+m-1$	55,56	(1),(2)
$H_{2,3}$	m	55	(1)
$H_{1,3}$	k	56	(2)
$H_{1,2}$	r	57	(3)
$H_{1,2,3}$	1	58	(0)

Table 7 should prove helpful in identifying the estimation equations and running a computer program correctly.

The tests for the various hypotheses are greatly facilitated by the following simple result:

Theorem 12: The LR-test statistic for any hypothesis H is given by

(61) $\quad Z(H) = 2 \sum_{ijl} X_{ijl} \{\ln X_{ijl} - \ln \hat{\mu}_{ijl}\},$

where $\hat{\mu}_{ijl}$ are the expected numbers under H. The LR-test statistic for H against the alternative H*, where H* $\longrightarrow$ H, is given by

(62) $\quad Z(H;H^*) = 2\Sigma\Sigma\Sigma X_{ijl}\{ln\hat{\mu}^*_{ijl}-ln\hat{\mu}_{ijl}\}$,

where $\hat{\mu}^*_{ijl}$ are the expected numbers under H*. $Z(H)$ is asymptotically χ^2-distributed with df=mkr-t, and $Z(H;H^*)$ is asymptotically χ^2-distributed with df=s-t, where

$\qquad$ S = number of canonical parameters under H*

and

$\qquad$ t = number of canonical parameters under H.

We note the existence of the simple formula.

$$Z(H;H^*) = Z(H)-Z(H^*) \; .$$

We can thus derive any test statistic of the form (62) from a table of Z(H) for all H involved. An example of such a table is shown in Table 8 below.

For any chain of successive tests, as, for example, (60) we can construct a table of variation as shown in Table 8, where 2.order interaction (1,2) means interaction between criterions 1 and 2.

Table 8: Table Of Variation For Successive Testing

Variation due to	Hypothesis	Test	Degrees Of Freedom
3. order interaction	H_{123}	$Z(H_{123})$	$(k-1)(r-1)(m-1)$
2. order interaction(1,2)	H_{12}	$Z(H_{12};H_{123})$	$(m-1)(k-1)$
2. order interaction(1,3)	$H_{12,13}$	$Z(H_{12,13};H_{12})$	$(m-1)(r-1)$
2. order interaction(2,3)	$H_{12,13,23}$	$Z(H_{12,13,23};H_{12,13})$	$(k-1)(r-1)$
Main effect (1)	H_1	$Z(H_1;H_{12,13,23})$	$m-1$
Main effect (2)	$H_{1,2}$	$Z(H_{1,2};H_1)$	$k-1$
Main effect (3)	$H_{1,2,3}$	$Z(H_{1,2,3};H_{1,2})$	$r-1$
Total		$Z(H_{1,2,3})$	$mkr-1$

Once more we stress the fact that we are performing successive tests. Hence, all previous hypotheses must be accepted before we proceed to the next hypothesis.

Example 3: We have earlier studied traffic data from Sweden as an

example of the multiplicative Poisson model. We now return to these
data. Data on accidents were collected in 18 weeks of 1961 and 18
weeks of 1962. In both years, a speed limit of 90 km per hour was
imposed on certain days. Accidents were recorded both on state
highways and on other roads. The observed accidents are thus categorized with respect to three criteria: type of road (1), speed limit
or not (2) and year (3). The observed numbers in this three-way contingency table are shown in Table 9.

Table 9: Number Of Traffic Accidents On Swedish Roads
In Parts Of 1961 And 1962 Classified according To Two
Criteria.

Year	Speed Limit	Highways	Other Roads	Total
1961	90 km	8	42	50
	Free	57	106	163
	Total	65	148	213
1962	90 km	11	37	48
	Free	45	69	114
	Total	56	106	162
Total	90 km	19	79	98
	Free	102	175	277
	Total	121	254	375

We start by obtaining the computer print-out of the Z-test statistics
for all of our basic hypotheses. Table 10 shows these quantities.

Table 10: Summary Of Test Statistics For Traffic Data

Hypothesis	$z(H)$	df
H_{123}	0.19	1
H_{12}	11.36	2
H_{13}	1.34	2
H_{23}	2.44	2
$H_{13,23}$	3.13	3

$H_{12,23}$	13.16	3
$H_{12,13}$	12.05	3
$H_{12,13,23}$	13.85	4
H_1	62.06	5
H_2	102.88	5
H_3	20.81	5
$H_{2,3}$	109.83	6
$H_{1,3}$	69.02	6
$H_{1,2}$	151.09	6
$H_{1,2,3}$	231.22	7

Table 10 suggests that we take the tests in the following successive order: $H_{123} \to H_{13} \to H_{13,23} \to H_{12,13,23} \to H_3 \to H_{1,3} \to H_{1,2,3} H_o$. The corresponding table of variation is shown in Table 11.

Table 11: Table Of Variation For Traffic Data

Variation Due To	Hypothesis	Test	Degrees Of Freedom
3rd order interaction	H_{123}	0.19	1
2nd order interaction (1,3)	H_{13}	1.15	1
2nd order interaction (2,3)	$H_{13,23}$	1.79	1
2nd order interaction (1,2)	$H_{12,13,23}$	10.72	1
Main effect (3)	H_3	6.96	1
Main effect (1)	$H_{2,3}$	48.21	1
Main effect (2)	$H_{1,2,3}$	162.20	1
Total		231.22	7

From Table 11 we see that H_{123}, H_{13} and $H_{13,23}$ can be accepted so that all interactions of 2nd and 3rd order are 0 except $\theta_{ij}^{(12)}$. By theorem 12, this means that criterion 3 is independent of criterion 1 as well as of criterion 2. Criterion 3 is the year. Hence, the analysis shows that the distribution of accidents according to road types is the same in 1961 and 1962, and that the ratio of accidents, with or without a speed limit, is the same in 1961 and 1962.

But there is still an interaction between road type and type of speed limit so that a speed limit influences accident numbers in a different way on different road types. The interaction estimates $\hat{\theta}_{ij}^{(12)}$ below show that a speed limit has more effect on highways than on other roads.

The main effect estimates $\hat{\theta}_i^{(1)}$ and $\hat{\theta}_j^{(2)}$ are difficult to interpret since we do not know the total length of the roads and since we have no information on the number of days with and without speed limits. The fact that H_3 is rejected is easier to interpret. Since the same number of weeks are recorded in 1961 and 1962, it follows that the number of accidents in 1961 is significantly larger than in 1962 whether there is speed limit imposed or not.

Under the hypothesis $H_{12,13}$, which is the last one to be accepted the sufficient marginals are $x_{ij.}$ and $x_{..l}$ for all i, j and l, as can be seen from Table 7. The estimates under $H_{13,23}$ are:

$\hat{\theta}_{ij}^{(12)}$	j = 1	2
i = 1	-0.22	+0.22
2	+0.22	-0.22

	i = 1	2
$\hat{\theta}_i^{(1)}$	-0.49	+0.49

	j = 1	2
$\hat{\theta}_j^{(2)}$	-0.62	+0.62

	l = 1	2
$\hat{\theta}_l^{(3)}$	+0.14	-0.14

$\hat{\theta}^{(o)} = 3.57.$

8. MULTIPLE CONTINGENCY TABLES

The general theory of multiple contingency tables is quite easily derived from our treatment of three-way tables. The notation becomes heavy, however, so we shall only sketch the general model and the various hypotheses and then indicate how it applies to 4-way tables. An M-way contingency table is a table of integers $x_{ij...tl}$ such that

$$\ln E[X_{ij...lt}] = \theta_{i...t}^{(1...M)} + \theta_{1...1}^{(1...M-1)} + ...$$

$$+\theta_{j..lt}^{(2...M)} + ...+\theta_{ij}^{(12)} +....+\theta_{1t}^{(M-1M)}$$

$$+\theta_i^{(1)} +....+\theta_t^{(M)} +\theta^{(o)} \quad ,$$

where all parameters sum to 0 over any index. The dimension of the table is $k_1 \cdot k_2 \cdot ... k_M$, where $i=1,...,k_1, j=1,...,k_2,...,l=1,k_{M-1}$ and $t=1,...,k_M$.

It is obvious that such a table contains an enormous number of parameters. Accordingly, it is a major problem simply to find a reasonable systematic way of treating the various parameters.

All parameters are termed <u>interactions</u> except for $\theta_i^{(1)},...,$ $\theta_t^{(M)}$ which are called <u>main effects</u> and $\theta^{(o)}$, which is termed the <u>over all effect.</u> We shall call

$$\theta_{ij...lt}^{(1...M)} \quad \text{an Mth order interaction}$$

$$\theta_{ij...l}^{(1...M-1)} \quad \text{an M-lth order interaction}$$

and so on. The likelihood function has kernal (the part containing both x's and θ's) equal to

$$\sum_i \sum_j ...\sum_l \sum_t x_{ij...lt} \left\{ \theta_{i...t}^{(1...M)} \right.$$

$$+\theta_{i...1}^{(1...M-1)} + ...+\theta_{ij}^{(12)} +...+\theta_i^{(1)} +...+\theta^{(o)} \left. \right\}$$

$$= \sum_i \sum_j ...\sum_l \sum_t x_{ij...lt} \theta_{i...t}^{(1...M)} +...$$

$$+ \sum_i \sum_j ...\sum_l x_{ij...l,} \theta_{i...1}^{(1...M-1)} +...$$

$$+\sum_i \sum_j x_{ij...,,} \theta_{ij}^{(12)} +...$$

$$+ \sum_i x_{i_\bullet\cdots_{\bullet\bullet}}\, \theta_i^{(1)} + \ldots + x_{\bullet\bullet\cdots_{\bullet\bullet}}\, \theta^{(0)}$$

Hence, all marginals of the table are minimal sufficient for the corresponding O's, and the O's are seen to be canonical parameters. Consequently, the likelihood equations take the form

$$x_{ij\ldots 1t} = E[x_{ij\ldots 1t}],$$

$$x_{ij\ldots 1_\bullet} = E[x_{ij\ldots 1_\bullet}],$$

$$\ldots$$

$$x_{ij\ldots_{\bullet\bullet}} = E[x_{ij\ldots_{\bullet\bullet}}],$$

$$\ldots$$

$$x_{i_\bullet\ldots_{\bullet\bullet}} = E[x_{i_\bullet\ldots_{\bullet\bullet}}]$$

$$\ldots$$

and

$$x_{\bullet\bullet\cdots_{\bullet\bullet}} = E[x_{\bullet\bullet\cdots_{\bullet\bullet}}].$$

We may then formulate various hypotheses in terms of θ's being 0. The ML-estimates for the parameters not set to 0 are obtained by solving the set of sufficient equations for the non-zero θ's. A test for H against H*, where more θ's are 0 under H than under H*, is then based on

$$(63) \quad Z(H;H^*) = \sum_i \sum_j \ldots \sum_1 \sum_t x_{ij\ldots 1t} \cdot \{\ln\hat{\mu}^*_{ij\ldots 1t} - \ln\hat{\mu}_{ij\ldots 1t}\},$$

where $\hat{\mu}_{ij\ldots 1t}$ are expected numbers under H* and $\hat{\mu}_{ij\ldots 1t}$ are the expected numbers under H. Normally, we try to find a sequence of H's such that any H implies all previous H's, and then test each H against the previous H* by (63). By theorem 3, $Z(H,H^*)$ is asymptotically χ^2-distributed with r degrees of freedom, where r = number of parameters under Π* - number of parameters under H.

All this was just intended as a brief summary of the elements of the analysis of an M-way table.

METHODS FOR THE ANALYSIS OF CONTINGENCY TABLES IN ROAD SAFETY RESEARCH.

S. Oppe

Institute for Road Safety Research SWOV,
Voorburg, The Netherlands

INTRODUCTION

A course on contingency table analysis for traffic
safety studies seems to be rather specialistic. To
investigate whether or not this subject matter is
worth to be selected as a subject for an ASI-meeting,
it is important to know the nature of the data in
traffic safety research. Moreover it is necessary to
know the power of the techniques for the analysis of
contingency tables (ct's) and their limitations. This
lecture is an attempt to give an outline of both
factors.
A ct, sometimes called a cross-table, is a table of
counts. Observations are classified according to one
or more characteristics. After the classification a
table is achieved, with numbers of observations in
each cell. In the analysis of such a table, one is
mainly concerned with the distribution of the
observations over the cells. The dominant aim in such
an analysis is the separation of the systematic
effects and the random fluctuations that together
resulted in the observed numbers. Therefore a model
is needed to describe the systematic effects and an
error theory that deals with the deviations of the
observed data from the data that are expected
according to the model.
The investigator is primarily interested in the
systematic effects. He wants to check assumptions

with regard to the underlying structure or model,
which process is called 'hypothesis testing', or he
wants to specify the underlying model structure, which
is called 'parameter estimation'. In both cases the
random component is important. The research worker
has to investigate to what extend random fluctuations
may have influenced his results. Therefore he needs a
theory that copes both with the underlying structure
and the nature of the random fluctuations.
As to this last aspect, much work has been done in the
past. With regard to the model specification, recent
research has resulted in many improvements that
account for an increased attention to the analysis of
ct's. The result of this research is also of great
importance for the analysis of traffic safety data.
Traffic safety is measured by traffic accidents.
Sometimes alternative measures are used such as the
number of traffic conflicts, when no accident data is
available. What follows regards to accident data and
conflict data, because both consist of counts. Even
when accident rates are used (accidents per head of
the population, per road length, per vehicle mile
travelled etc.) a measure of traffic safety results
which is in fact based on counts.
If one concentrates on the fact that these measures
result from counts, then the analysis will tend to an
explanation of that number of counts in that
particular cell of the table. This leads to questions
like: 'what is the probability of that number of
observations in that particular cell?' A completely
different approach is got if one stresses the point
that he deals with a measure of safety, a safety
score. The relevant question then seems to be: 'what
makes the score for that particular cell so high or so
low?' Both types of analysis do appear in traffic
safety research.
The analysis of ct's is primarily concerned with the
distribution of observations over cells. The second
approach leads in most cases to an analysis in the
context of linear models, such as linear regression or
the analysis of variance. The approaches however lead
to completely different model assumptions. In the
analysis of variance approach the expected value for
the score in cell $<i,j>$ of a two-way table, is
generally supposed to result from two additive
components, according to the row and column position
of the cell:

$$E(x_{ij}) = r_i + c_j$$

In the analysis of ct's a multiplicative model is generally assumed, although additive models do exist. In many practical cases the choice of the model structure depends less on the nature of the data and more on factors such as the statistical complexity of the model, or the knowledge about or availability of the techniques. From a statistical point of view, the linear model has many advantages, resulting from the applicability of the theory of linear vector spaces on statistics and the statistical properties related with linear transformations. The presentation of the multiplicative model for the analysis of ct's as a linear model for the log-counts resulted in the same advantages and largely accounts for the increased interest in the ct-analysis.
Before we go into more detail with regard to the model structure it seems good to recollect some of the basic theories of ct-analysis.

THE MULTINOMIAL MODEL

In probability theory there are a number of so called 'ideal experiments' that may be used as a model for the analysis of ct's. One of these theoretical experiments leads to the multinomial model. The experiment consists of a number of repeated independent draws of a sample element from a population. Each draw resulting in one of several possible outcomes. To apply this model, the following assumptions are needed:
- the probability that a particular observation is classified into a certain cell is independent from the classification of other observations
- the probability is the same for each observation
- the observations are classified in one and only one cell, in other words the events are mutually exclusive and exhaustive.
- no assumptions are made about how the events occur. The occurrence of events is taken for granted.
These assumptions lead to the following probability distribution: The probability that from a number n of observations, x_i observations are classified in class i, is equal to:

$$P\ (X_1 = x_1 \wedge \dots \wedge X_i = x_i \wedge \dots \wedge X_m = x_m) = n! \prod_i^m \frac{1}{x_i!}\ p_i^{x_i}$$

where p_i is the probability for each observation to be classified in class i. The expected number of observations in each class is equal to $n.p_i$; the variance is equal to $n.p_i.(1-p_i)$ and the covariance between the observations in class i and class j is equal to $-n.p_i.p_j$.

For traffic safety research it may be concluded that the model can be used for the analysis of ct's that consist of accident numbers. However, if the observations are numbers of cars involved in accidents or numbers of persons injured in accidents, then the model is not applicable because many observations are collected in groups and therefore not independent anymore.

For the analysis of ct's it is important to realise that the events are in fact composite events. The classification of an observation in cell <i,j> means a classification in row i and column j. This results in a special restriction of the model. It is often assumed that the classification of an observation in row i is independent from the column where the observation is in and vice versa. Two events A and B are said to be statistically independent if the probability $P(A \cap B)$ that both events occur is equal to the product $P(A).P(B)$ of the probabilities that each of the events occurs. The independence of an observation in row i from being in column j is therefore expressed in the model with the restriction that the probability p_{ij} of an observation in cell <i,j> is the product $p_i.p_j$ of the probabilities of an observation in row i and in column j.

Thus the hypothesis of no interaction between rows and columns leads to a multiplicative model for the cell counts. The common used Chi-square test of no interaction results from this model. Here the probabilities p_i and p_j are estimated from the marginal row and column distributions of the table.

THE POISSON MODEL

Apart from the multinomial model there is another model that is in use in the analysis of ct's with traffic safety data. This is the Poisson model.

For the application of this model all the assumptions of the multinomial model are needed together with one extra assumption regarding the occurrence of observations. Here the observations are not taken for granted, but it is supposed that the occurrences are

Poisson distributed with parameter λ. The probability distribution is written as follows:

$$P(k;\lambda) = e^{-\lambda} . \lambda^k /k!$$

which means that the probability of exactly k observations, given the Poisson parameter λ, is equal to the expression at the right side of the equation. If we make this assumption and futher assume that the observations are multinomially distributed over the cells of the table with probabilities p_{ij}, then it is proved that the distribution of the number of observations in each cell is Poisson distributed with probability λp_{ij}. Otherwise it can be proved that if the number of observations in each cell is Poisson distributed with Poisson parameter λp_{ij}, then the total number of observations is also Poisson distributed with parameter $\lambda = \sum \lambda p_{ij}$.

Moreover it is easily proved that the conditional distribution of the observations over the cells, given the total number of observations, is a multinomial distribution with probabilities p_{ij}. In formula:

$$P\left(X_1 = x_1 \wedge \ldots \wedge X_m = x_m \,\middle|\, \sum_i^m x_i = k\right) =$$
$$= \prod_i^m \left\{ e^{-\lambda p_i} (\lambda p_i)^{x_i} / x_i! \right\} \Big/ \left\{ e^{-\sum \lambda p_i} (\sum \lambda p_i)^k / k! \right\} =$$
$$= k! \prod_i^m \frac{1}{x_i!} p_i^{x_i}.$$

Thus, in this context, the multinomial distribution can be regarded as a restricted case of the Poisson model.

To investigate in which cases the Poisson assumption is satisfied, it is useful to start with the most common interpretation of the Poisson model.

If we apply the before mentioned theoretical experiment to the occurrence of observations, then a trial can be regarded as a unit period of time during which an event may or may not occur with probability p (more than one event may occur). If the time period is divided in n equal parts then we assume for each period of time that the probability of an event is p_m =p/n. This results in n trials each with probability p_n of an event to occur. The expected number of total events remains the same, being equal to $\lambda = n.p_n$.

If n tends to infinity, then the probability of more than one event becomes negligible and the trials may

be interpreted as independent binomial trials.
It is proved that the limit distribution of the total
number of events in this case is equal to the Poisson
distribution.
The mean and variance of the Poisson distribution both
equal λ .
The essential assumptions are that the occurrence of
an event does not depend on the history of previous
trials, or in other words that the trials are
independent and that the probability of an event is
equal for each trial, or that the events occur
homogeneous in time. The first assumption does not
find much resistance in traffic safety research:
accidents are rare events and seldom one accident
causes another. In most cases where this however is
true, one often regards such a chain of accidents as
one (complicated) accident. The second assumption
however troubles many investigators. Traffic flow
changes rapidly over time and the accident rate is
supposed to change with it. However, the homogeneity
assumption needs not to hold over a long period of
time. From the fact that the sum of Poisson
distributed variables is again Poisson distributed, it
follows that it is enough that the homogeneity
assumption holds for a short period. In many cases
support is found for the Poisson assumptions. In
cases where these assumptions does not hold, one
sometimes assumes a mixed or compound Poisson process.
In these cases the variance exceeds the mean, which is
true for instance with the negative binomial
distribution. Here it is assumed that the sampling is
from Poisson distributions with different parameters.
The statistical properties of these compound
distributions lead to serious complications as far as
the analysis of ct's is concerned and therefore do not
lead to practical alternatives.
The Poisson distribution is used in many
investigations of traffic safety research. Recently
the multiplicative Poisson model has been used for the
analysis of ct's with regard to accident data.
Rasch(1973) applied the model to accidents classified
according to road categories and days. Furthermore he
used the model to estimate parameters for accident
proneness of different drivers and to test wether or
not these parameters changed with time. For these
test he used the Chi-square test, based on the
conditional Poisson distributions.
Hamerslag(1977) uses the multiplicative Poisson model
to estimate the parameters for different classes of
several accident characteristics jointly, under the

hypothesis of independence between the characteristics. Which is a rather strong assumption, that only can be released by combining variables in one new variable.

De Leeuw & Oppe (1976) used a weighted version of the Poisson model. It has been applied for instance to ct's with accident numbers collected over different areas, or periods of time. This model is rather similar to the Multiplicative Poisson model with unequal cell rates of Andersen (1977).

RECENT DEVELOPMENTS IN CT-ANALYSIS

We shall now come to a more systematic description of the recent developments.

Most of the applications of an analysis of ct's are instances of testing the hypothesis of no interaction in two-way tables. Probabilities of row and column classification are estimated from the row and column marginals.

From these values the expected number E_{ij} of observations in each cell is computed as $n.p_i .p_j$ and compared to the observed number of counts O_{ij} .

A measure of discrepancy X^2 between both series of values called Chi-square is defined as:

$$X^2 = \sum_{i,j} (O_{ij} - E_{ij})^2 / E_{ij} .$$

Under the assumption that the hypothesis of no interaction is true, the value of X^2 depends only on random fluctuations. The distribution of X^2 is therefore assumed to be based on the properties of the multinomial distribution only.

For each total number of observations and each model specified in terms of cell-probabilities, there is a discrete set of possible values of X^2, the probability of each value depending on the probability of the corresponding set of cell-observations, given that specified multinomial model. Computation of these exact probabilities is rather cumbersome. Only in very restricted cases tests based on these exact distributions of X^2 are of practical use. Fishers exact test for 2x2 tables is an instance of these tests. Therefore , in practice additional assumptions are made in order to arrive at the distribution of X^2-values. In fact it is assumed then that the value of X^2 is distributed as the sum of a number of squared standard normal variates. This results from a well known limit theorem for the multinomial distribution. Practical values of this theoretical distribution denoted with χ^2 and known as the Chi-square distribution, can easily be found from tables if the number of squared independent standard normal variates is known. This number is often called the degrees of freedom (df). The use of these tables is only warranted when relatively large numbers of observations are present. Therefore one often speaks of large sample tests.

Not much is known about the usefullness of the χ^2-distribution in small samples, other than with 2 * 2 tables. There are a number of handrules for usefullness in some situations. An overview can be found in Cochran (1952).

Oppe(forthcoming) investigates the behaviour of maximum likelihood and modified minimum Chi-square estimates for log-linear parameters and related X^2-distributions for a number of tables with small expected cell counts. This is done by means of the Monte Carlo method.

The necessity of large samples often brings investigators in a rather difficult position. In many cases there are only small numbers of accidents on which statistical analysis can be based. This is one of the reasons why the Chi-square test did not get much attention for a long time. There is another reason. The Chi-squared test as described above is used as a test of no interaction. If this hypothesis has to be rejected, then the test does not tell us in what way the model fails to describe the data, but only that it fails to do so. In other words, the Chi-square analysis is a poor instrument for theory building.

Another problem often mentioned in hypothesis testing, using the Chi- square test, is the fact that small and uninteresting deviances from the null-hypothesis will lead to a rejection of that hypothesis when a very large number of data has been collected. Thus besides the question of significance, there is the question of relevance. This problem stresses the need for parameter estimation and computation of their confidence regions as information additional to hypothesis testing.

The reason why the interest in the analysis of ct's is increased specially since 1960 is not because the problem of small numbers is solved but mainly because of a more detailed model testing procedure and the

increased possibilities for parameter estimation.
There are four main developments that must be
mentioned:
1. The application of the Chi-square test as a test
for the hypothesis of no interaction, is generalised
to ct's with more than two ways of classification.
Foldvary & Lane (1974) in measuring the effect of
compulsory wearing of seat belts, used the method of
partitioning the total Chi-square in Chi-square values
for first, second and third order interactions in a
four-way table.
2. Test are not restricted to overall effects, but
Chi-square values are decomposed with regard to
subhypotheses of the model. Not only with regard to
the different levels of interaction between variables
as used by Foldvary & Lane, but also within for
instance the two-way table to partition the total
Chi-square in Chi-squares with respect to (subgroups)
of single classes. The advantage over earlier methods
where the subtables are analysed as such, is that the
partitioning of the total Chi-square is exact and
results in independent Chi-squares for subhypotheses.
3. As a result of this last mentioned development one
has to put more attention to parameter estimation
also. Subhypotheses as mentioned are derived from
constraints on the parameters in the model, such as
linear constraints, quadratic constraints on the
row-parameters etc.
4. The unit of analysis has been widened from counts
to weighted counts, where the counts are weighted
before analysis. The hypothesis testing of main
effects is sometimes possible with weighted numbers,
for instance if one wants to test wether or not there
is a significant difference between the number of
fatalities in different countries per head of the
population.
However the most important reason for using these
extensions of the Chi-square method comes from the
presentation of the theory in terms of linear models.
The linear model is assumed to exist for the
log-counts. Therefore it has been called a log-linear
model.
The log-linear model states that the logarithm of the
expected value of the cell counts can be decomposed as
follows:

$$\ln(E(x_{ij})) = \mu + \alpha_i + \beta_j + \gamma_{ij} .$$

If the parameters are not known, and must be estimated
from the data, then it is always possible to find a
perfect solution for the parameters of the above
stated (saturated) model. Testing independence of
rows and columns means that the hypothesis $\gamma_{ij} = 0$
$\gamma_{i \cdot j}$ is tested.
Stricter models such as $\ln(E(x_{ij})) = \mu + \alpha_i$ can be
tested within the former model. It is also possible
to test restrictions on parameters, such as linearity
restrictions for instance on the α's.
Moreover, the generalisation to higher order tables is
going straight forward and tests of restricted models
are conceptually clear. We shall not go into more
detail now. A comprehensive description is given in
Bishop, Fienberg & Holland (1975).
Sometimes the log-linear model is confused with the
log-normal model. The log-normal model is used in the
analysis of variance to stabilise the variance of the
observations for all kinds of measurement.
The strong resemblance between both techniques is best
illustrated in Nelder & Wedderburn (1972). In their
computerprogram GLIM they incorporated the log-linear
model for the analysis of counts as a special case of
the linear model. Only the log-transformation of the
data is needed to apply the linear analysis.
This leads us back to the question about the additive
and multiplicative interaction in ct's. Darroch(1974)
who compares both models from a statistical point of
view, gives the following representation of no
interaction in a three-way table:
For the multiplicative model; $p_{ijk} = \alpha_{ij} \cdot \beta_{ik} \cdot \gamma_{jk}$
and for the additive model; $p_{ijk}/(p_i \cdot p_j \cdot p_k) =$
$\alpha_{ij} + \beta_{ik} + \gamma_{jk}$.
This additive model was first introduced by
Lancaster(1951). He used the model to partition the
total Chi-square (that can be regarded as a measure of
residuals) according to the α- and β- and γ-values.
In this way the residuals are tested futher according
to an additive model for row and column and layer
effects. This kind of a compromise between
interaction seen as statistical independence and a
decomposition of the residuals according to an
additive interaction model leads to different results
in tables of higher order than two-way tables.
Despite some advantages of the additive interaction
model, Darroch concludes to a slight preference of the
multiplicative model based on the statistical
properties only.
Oppe (forthcoming) applies a generalised linear model
to analise a table of accident rates according to wet

pavement conditions and hourly traffic volume classes. He uses an additive model: $E(a_{ij}^*) = r_i + c_j$. This model however was not applied to the datapoints a_{ij} themselves. A monotone transformation a_{ij}^* of the data was used that led to the best fit of the model. A description of this kind of monotone regression models is found in Kruskal (1965).

The transformation turned out to be logarithmic, suggesting a multiplicative model rather then an additive model. Therefore there are, besides the logical attractiveness of the multiplicative model and some statistical advantages, also emperical results that support the multiplicative model for this kind of accident data. In many cases these tests of the model are ignored and a poor fit, resulting from the use of an incorrect model is much too lightly interpreted as just random error. For instance, in linear regression models one seldom is interested in the magnitude of the error component. In log-linear tests, however, this is not the case. The test of the model implies assumptions about the magnitude of the error component which is a great advantage of the technique. It leads to a quicker rejection of an incorrect model.

REFERENCES

1. Andersen, E.B. (1977). Multiplicative Poisson models with unequal cell rates. Scand. J. Statist. 4.

2. Bishop, Y.M.M., Fienberg, S.E. & Holland, P.W. (1975). Discrete Multivariate Analysis; Theory and Practice. MIT-Press, London, 1975.

3. Darroch, J.N. (1974). Multiplicative and additive interaction in contingency tables. Biometrika, 1974, p. 207.

4. De Leeuw, J. & Oppe, S. (1976). The analysis of contingency tables; Log-linear Poisson models for weighted numbers. SWOV, Voorburg, 1976.

5. Foldvary, L.A. & Lane, J.C. (1974). The effectiveness of compulsory wearing of seat belts in

casualty reduction.   A.A.P., 6, p. 59-81.

6. Hamerslag, R. (1977). Het gebruik van het multi-proportionele schattingsmodel bij ongevallenanalyse. Rapport van ingenieursbureau Dwars, Heederik en Verhey, Amersfoort, Nederland.

7. Kruskal, J.B. (1965). Analysis of factorial experiments by estimating monotone transformations of the data. J. of the R., Sta. Soc., Serie B, 27, 1965.

8. Nelder, J.A. & Wedderburn, R.W.M. (1972). Generalized Linear Models. J.R. Statist. Soc. A, 1972, p. 370.

9. Rasch, G. (1973). Two applications of the multiplicative Poisson models in road accidents statistics. In: Proc. of the 38th Session of the ISI, Wien, 1973.

STATISTICAL TESTS ON POISSON VARIABLES FOR ROAD SAFETY
EVALUATION

S. Lassarre

Organisme National de Sécurité Routière,
Arcueil, France.

ABSTRACT

The comparison of two or four Poisson variables arises very
often in effectiveness evaluation for Road Safety.
By applying the theory of conditional tests, we can construct
UMP unbiased tests for different cases: one-sided or two-sided
hypotheses and small or large samples.

1. INTRODUCTION

In effectiveness evaluation for road safety studies, we usually make two sorts of experiments:

— One experiment on an experimental area or set of roads or
intersections which is observed before and after the treatment;

— Another experiment with an experimental and a control group
which are observed before and after the treatment.

The comparison before and after treatment is done according to
the observed number of accidents which is supposed to be a Poisson
variable. The classical hypothesis to be tested in an evaluation
study is that the treatment or the countermeasure has produced a
reduction in the number of accidents between the two periods. The
most adapted hypothesis is a one-sided hypothesis such as

$H_1: \xi_1 = \xi_2$ against an alternative $K_1: \xi_1 > \xi_2$,

which means that we want to test an equality against a decrease in
the number of accidents between the two periods. We can also try a

two-sided hypothesis such as

H_2: $\xi_1 = \xi_2$ against K_2: $\xi_1 \neq \xi_2$,

where we test an equality against a difference in the number of accidents. In this case, we don't take care of the a priori known and expected effect of a countermeasure which is a reduction of accident. These two hypotheses are composite because they include several parameters.

The comparison of two or four Poisson variables is a direct application of the theory of UMP unbiased tests for multiparameter families. The procedure of randomized tests extracted from Lehmann's book [8] will help us to build tests for these comparisons in the case of one-sided and two-sided hypotheses, paying attention to small and large samples.

2. UMP UNBIASED TESTS FOR MULTIPARAMETER EXPONENTIAL FAMILIES

Let X be a random variable distributed according to an exponential family:

$$dP^X_{\vartheta,\lambda}(x) = C(\vartheta,\lambda)\exp[\vartheta U(x) + \sum_{i=1}^{k} \lambda_i T_i(x)]d\mu(x),$$

with $(\vartheta,\lambda) \in \Omega$ a convex parameter space with dimension $k + 1$,

ϑ a real-valued parameter on which the hypothesis is made,

$\lambda = (\lambda_1...\lambda_k)$ k remaining parameters occurring as nuisance parameters,

$T = (T_1,...,T_n)$ a k vector of real-valued functions.

We consider two hypotheses H_j and alternatives K_j

H_1: $\vartheta \geqslant \vartheta_0$ (or $\vartheta = \vartheta_0$), K_1: $\vartheta < \vartheta_0$,

is left unspecified

H_2: $\vartheta = \vartheta_0$, K_2: $\vartheta \neq \vartheta_0$.

Because of the exponential form of the distribution, the optimal test is expressed as a function of the sufficient statistics U and T.

We can limit our attention to the joint distribution of sufficient statistics (U,T)

$$dP^{U,T}_{\vartheta,\lambda}(u,t) = C(\vartheta,\lambda)\exp[\vartheta u + \sum_{i=1}^{k} \lambda_i t_i]d\nu(u,t).$$

The conditional distribution of U given $T = t$ has an exponential form

$$dP_{\vartheta}^{U,T}(u) = C_t(\vartheta)\exp(\vartheta u)d\nu_t(u).$$

Because this distribution no longer depends on the parameter λ, the conditional problem takes a simple and classical form. By a corollary of the Neyman-Pearson Lemma, in the conditional situation, there exists a UMP test with critical function φ_1 satisfying

$$\varphi_1(u,t) = \begin{cases} 1 & \text{when } u < C_0(t), \\ \gamma_0(t) & \text{when } u = C_0(t), \\ 0 & \text{when } u > C_0(t), \end{cases}$$

where the functions C_0 and γ_0 are determined by

$$E_{\vartheta_0}[\varphi_1(U,T)/t] = \alpha \text{ for all } t.$$

There exists also an unbiased UMP test for H_2 with critical function φ_2 satisfying

$$\varphi_2(u,t) = \begin{cases} 1 & \text{when } u < C_1(t) \text{ or } u > C_2(t), \\ \gamma_i(t) & \text{when } u = C_i(t), \ i = 1,2, \\ 0 & \text{when } C_1(t) < u < C_2(t). \end{cases}$$

The C's and γ's are determined by

$$E_{\vartheta_0}[\varphi_2(U,T)/t] = \alpha,$$

$$E_{\vartheta_0}[U\varphi_2(U,T)/t] = \alpha E_{\vartheta_0}[U/t].$$

The critical functions φ_1 and φ_2 characterize a randomized test. For any value of u, a randomized test chooses between two decisions, rejection and acceptance of the null hypothesis, with two probabilities that depend on u and denoted by $\varphi(u)$ and $1 - \varphi(u)$. (For a non-randomized test, the values of $\varphi(u)$ are 1 or 0). It a value of u occurs, a random experiment is performed with probabilities $\varphi(u)$ and $1 - \varphi(u)$ on which the decision is based.

If we examine the first critial function, we have to take the following decisions:

if $u < C_0(t)$ then $\varphi_1(u,t) = 1$, we reject the hypothesis H_1,

if $u > C_0(t)$ then $\varphi_1(u,t) = 0$, we accept the hypothesis H_1, if $u = C_0(t)$ then $\varphi_1(u,t) = \gamma_0(t)$, we accept or reject the hypothesis H_1 after an auxiliary randomization with a reject probability equal to $\gamma_0(t)$.

The condition $E_{\vartheta_0}(\varphi_1(u,t)/t) = \alpha$ means that the conditional probability of rejection of H_1 when it is true ($\vartheta = \vartheta_\vartheta$) must be equal to the predetermined size of the test α.

For the second critical function, two critical regions are defined by $C_1(t)$ and $C_2(t)$ in relation to the two-sided hypothesis. The additive equality $E_{\vartheta_0}[U\varphi_2(U,T)/t] = \alpha E_{\vartheta_0}(u/t)$ comes from unbiased properties and a condition on the minimum of the power function for $\vartheta = \vartheta_0$.

So far, the critical functions φ_j have been considered as conditional tests given $T = t$. As the statistic T has the good properties (sufficiency for λ, completeness for the null hypothesis) these conditional unbiased tests are also unconditional unbiased tests. This is an important theorem stating that with the properties of the statistic T of nuisance parameters we can solve the unconditional test problem by treating the conditional test problem which is the solution of the classical Neyman-Pearson Lemma. So the critical functions φ_1 and φ_2 as defined above constitute UMP unbiased tests for testing hypotheses H_1 and H_2 in the case of a multiparameter exponential family [8,10].

3.   COMPARISON OF TWO POISSON DISTRIBUTIONS

Let X and Y be two variables independently distributed according to two Poisson distributions of parameters λ and μ. Their joint distribution

$$P(X = x, Y = y) = \frac{e^{-(\lambda+\mu)}}{x!y!} \exp[y\log\frac{\mu}{\lambda} + (x + y)\log\lambda]$$

belongs to an exponential family with $\rho = \log\mu/\lambda$, $U = Y$, $T = X + Y$. We can test two composite hypotheses (cf., Introduction) [8]:

H_1: $\mu \geqslant \lambda$ (or $\mu = \lambda$), K_1: $\mu < \lambda$,

H_2: $\mu = \lambda$, K_2: $\mu \neq \lambda$,

which can be translated with $\vartheta = \mu/\lambda$, as

H_1: $\vartheta \geqslant 1$ (or $\vartheta = 1$), λ, K_1: $\vartheta < 1, \lambda$,

H_2: $\vartheta = 1, \lambda$, K_2: $\vartheta \neq 1, \lambda$,

Comparing with the formula above, we have $U = Y$ and $T = X + Y$.

So by applying the preceding theorem there exists UMP unbiased
tests of these hypotheses which are performed conditionally on the
integer line $X + Y = t$.
The conditional distribution of Y given $T = t$,

$$P(Y = y, X + Y = t) = \binom{t}{y} \left(\frac{\mu}{\lambda + \mu}\right)^y \left(\frac{\lambda}{\lambda + \mu}\right)^{t-y},$$

$$y = 0,1,\ldots,t,$$

is a binomial distribution corresponding to t trials with probability $p = \frac{\mu}{\lambda + \mu}$ of success.

The conditional problem becomes a problem of testing the parameter p of a binomial law concerning the conditional Y. The null
hypothesis H: $\mu \geqslant a\lambda$ becomes H: $p \geqslant a/(a+1) = p_0$ which is rejected
when Y is too small. The cut-off point depends on addition to a,
on t. To test the equality of two Poisson distributions, we take
$p_0 = 1/2$. When the periods of observation are different before
and after the treatment, we take p_0 as the ratio of the after
period length to the total period of observation. When a comple-
mentary of information as an evolution estimate C on a control
area is supplied, we can take $p_0 = \frac{C}{1 + C}$ (Tanner [9]).

3.1. One-Sided Hypothesis

$$H_1: p \geqslant p_0 \text{ (or } p = p_0), \quad K_1: p < p_0.$$

3.1.1. Small Sample.

The UMP unbiased test is determined by the following critical
function:

$$\varphi_1(y,t) = \begin{cases} 1 & \text{when } y < C_0(t), \\ \gamma_0(t) & \text{when } y = C_0(t), \\ 0 & \text{when } y > C_0(t), \end{cases}$$

with the $C_0(t)$ and $\gamma_0(t)$ determined by the formula:

$$\sum_{y=0}^{C_0(t)-1} \binom{y}{t} p_0^y (1 - p_0)^{t-y} + \gamma_0(t) \binom{t}{C_0(t)} p_0^y (1 - p_0)^{t-y} = \alpha.$$

The solution valid for small samples can be found with the aid

of statistical tables. Practically, we fix the size of the test
at a level α (equal to 0.05 usually). We select a number $C_0(t)$
for which $P(Y \leqslant C_0(t) - 1) < \alpha$ and $P(Y \leqslant C_0(t)) > \alpha$ by the aid of
statistical tables. If we take the cut-off at $C_0(t) - 1$ or $C_0(t)$,
the size will not be equal to α. So we have to adjust the test
size by an auxiliary randomization in calculating the probability
$\gamma_0(t)$ such as:

$$P(Y \leqslant C_0(t) - 1) + \gamma_0(t)P((Y = C_0(t)) = \alpha.$$

The process of decision is now ready:

if $y < C_0(t)$ we reject the hypothesis,

if $y > C_0(t)$ we accept the hypothesis,

if $y = C_0(t)$ we make an auxiliary random experiment with probability of reject equal to $\gamma_0(t)$ to decide whether to accept
or reject the hypothesis.

Example 3.1.1

The number of accidents at an intersection in two years comes
from 17 to 8 after an improvement. The two periods of observation
are equal, so we take $p_0 = 1/2$. We have $t = x + y = 25$. To find
the cut-off $C_0(t)$, we calculate the different probabilities of
$y = 0,1,\ldots,t$ for a binomial law with parameters $p_0 = 1/2$ and
$t = 25$ and sum these probabilities up to the critical level α
equal to 0.005.

y	Probability $B(\frac{1}{2},25)$ $P(Y = y)$	Cumulative Probability $B(\frac{1}{2},25)$ $P(Y \leqslant y)$
0	0.00000	0.00000
1	0.00000	0.00000
2	0.00001	0.00001
3	0.00007	0.00008
4	0.00038	0.00046
5	0.00158	0.00204
6	0.00528	0.00732
7	0.01433	0.02165
8	0.03223	0.05388
9	0.06089	0.11477

For a critical level α equal to 0.05, we have to take $C_0(t) = 8$ and

$$\gamma_0(t) = \frac{0.05 - P(Y \leqslant 7)}{P(Y = 8)} = \frac{0.05 - 0.02165}{0.03223} = 0.88.$$

The critical function becomes:

$$\varphi(y, 26) = \begin{cases} 1 & \text{if } y < 8 \text{ we reject the hypothesis,} \\ 0 & \text{if } y > 8 \text{ we accept the hypothesis,} \\ 0.88 & \text{if } y = 8 \text{ we make a random experiment with} \\ & \text{probability of rejection equal to} \\ & 0.88 \text{ to decide whether to accept} \\ & \text{or reject the hypothesis.} \end{cases}$$

For our problem $y = 8$, so we have to make this random experiment to choose between the two hypotheses. But, as the probability of rejection is 0.88, we have a greater chance to conclude the rejection of the hypothesis $p \geqslant p_0$, that is to say to conclude a significant reduction of accidents due to the countermeasure.

3.1.2 Large Sample

The binomial law can be approximated by a normal law $N(tp_0, tp_0q_0)$. Using the standardized variable $\frac{y - tp_0}{\sqrt{tp_0q_0}}$ which tends to the normal distribution $N(0,1)$ and u_α, the $1 - \alpha$ percentile of the Gaussian law $N(0,1)$, the critical function defining an UMP unbiased test becomes:

$$\varphi(y, t) = \begin{cases} 1 & \text{when } y < tp_0 - u_\alpha\sqrt{tp_0q_0}, \\ 0 & \text{when } y > tp_0 - u_\alpha\sqrt{tp_0q_0}. \end{cases}$$

Example 3.1.2

From a survey of roads with and without plantations, two and a half years of data collection give the following numbers of accidents: 1,487 on roads with plantations, 1,142 on roads without plantations. To take care of the different lengths and traffic flows in these two samples, we calculate a corrective factor:

$$k = \frac{\text{number of vehicles} \times \text{km for roads with plantations}}{\text{number of vehicles} \times \text{km for roads without plantations}}$$

The value of k is 0.80, which translates the fact that the two samples have been carefully matched. We take as p_0 instead of $\frac{1}{2}$, $\frac{k}{1 + k}$ = 0.44 with x = 1,487, y = 1,142, t = 2,629.

For a critical level equal to 0.05, u_α = 1.64 and:

$$C_0(t) = 2,629 \times 0.44 \ \times 1.65\sqrt{2,629 \times 0.44 \times 0.56}$$

$$= 1,114.7.$$

We can define the critical function

$$\varphi(y,2,629) = \begin{cases} 1 \text{ if } y < 1,115 \text{ we reject the hypothesis,} \\ 0 \text{ if } y > 1,115 \text{ we accept the hypothesis.} \end{cases}$$

In our problem y = 1,142, so we accept the hypothesis of no effect between the number of accidents on roads with or without plantations.

3.1.3 Correction for Continuity

For discrete distribution, the probability at value a is taken to refer to the interval $[a - \frac{1}{2}, a + \frac{1}{2}]$. The probability $P(A \leqslant a)$ of a discrete variable A must be taken at $a + \frac{1}{2}$ and approximated by $p' = \int_0^{a+\frac{1}{2}} f(A)\,dA$ where $f(A)$ is the Gaussian density function with parameters as derived from the binomial distribution [3].

The critical function becomes

$$\varphi(y,t) = \begin{cases} 1 \text{ when } y < tp_0 - \frac{1}{2} - u_\alpha \ tp_0q_0, \\ 0 \text{ when } y > tp_0 - \frac{1}{2} - u_\alpha \ tp_0q_0, \end{cases}$$

when we take as approximation $p'' = \int_0^a f(A)\,dA$ as done before in 3.1.2, we approximate the cumulative probability $P'' = \frac{1}{2}P(A = a) + P(A \leqslant a - 1)$ which is also a good approximation of $P(A \leqslant a)$.

The two critical functions with or without continuity correction can receive theoretical justifications and both of them can be employed.

3.2 Two-Sided Hypothesis

$$H_2: p = p_0, \qquad\qquad K_2: p \neq p_0.$$

3.2.1 Small Samples

The UMP unbiased test is determined with the following critical function

$$
\varphi_2(y,t) = \begin{cases} 1 & \text{when } y > C_2(t) \text{ or } y < C_1(t), \\ \gamma_i(t) & \text{when } y = C_i(t), \ i = 1,2, \\ 0 & \text{when } C_1(t) < y < C_2(t). \end{cases}
$$

$C_i(t)$ and $\gamma_i(t)$ can be found by resolving two equations with the
aid of statistical tables results. The problem takes a simpler
form, for samples which are not too small, and values of p_0 not
too close to 0 or 1. The distribution of Y is approximately sym-
metric with respect to the origin. We can use a simple "equal
tails test" as approximate with c's and γ's defined by:

$$
\sum_{y=0}^{C_1(t)-1} \binom{t}{y} p_0^y q_0^{t-y} + \gamma_1 \binom{t}{C_1(t)} p_0^{C_1(t)} q_0^{t-C_1(t)}
$$

$$
= \gamma_2 \binom{t}{C_2(t)} p_0^{C_2(t)} q_0^{t-C_2(t)} + \sum_{y=C_2(t)+1}^{t} \binom{t}{y} p_0^y q_0^{t-y}
$$

$$
= \alpha/2.
$$

Example 3.2.1

We take again $x = 17$ and $y = 8$ accidents before and after improvement and construct an equal-tail test of critical level α.
We define two cut-offs $C_1(t)$ and $C_2(t)$ by examining the cumulative
probabilities round the value $\alpha/2$ equal to 0.025 by $C_1(t) = 8$,
$C_2(t) = 17$. The probabilities of rejection are:

$$
\gamma_1(t) = \gamma_2(t) = \frac{0.025 - 0.02165}{0.05388} = 0.06
$$

The critical function can be defined as:

$$
\varphi(y,26) = \begin{cases} 1 & \text{if } y < 8 \text{ or } y > 17 \text{ we reject the hypothesis,} \\ 0 & \text{if } 8 < y < 17 \quad \text{we accept the hypothesis,} \\ 0.06 & \text{if } y = 8 \text{ or } y = 17 \text{ we make a random experi-} \\ & \qquad \text{ment with a probability} \\ & \qquad \text{of rejection equal to 0.06.} \end{cases}
$$

The actual value y is 8; so to decide the acceptance or the rejection we have to make a random experiment with probability of rejection equal to 0.06. In this case, we will rather conclude that we accept the hypothesis of no difference between accidents before and after treatment.

3.2.2 Large Samples

When n is sufficiently large, the constants can be directly determined from the normal distribution. We can define the critical function with:

$$
\varphi(y,t) = \begin{cases} 1 \text{ when } \begin{array}{l} y < tp_0 - u_{\alpha/2}\sqrt{tp_0 q_0} \text{ or} \\ y > tp_0 + u_{\alpha/2}\sqrt{tp_0 q_0}, \end{array} \\ 0 \text{ when } tp_0 - u_{\alpha/2}\sqrt{tp_0 q_0} < y < tp_0 + u_{\alpha/2}\sqrt{tp_0 q_0}. \end{cases}
$$

We can also introduce a continuity correction for the one-sided test by subtracting or adding 1/2 to the estimated mean tp_0. The critical function becomes:

$$
\varphi(y,t) = \begin{cases} 1 \text{ when } \begin{array}{l} y < tp_0 - \frac{1}{2} - u_{\alpha/2}\sqrt{tp_0 q_0} \\ y > tp_0 + \frac{1}{2} + u_{\alpha/2}\sqrt{tp_0 q_0} \end{array} \\ 0 \text{ when } tp_0 - \frac{1}{2} - u_{\alpha/2} tp_0 q_0 < y < tp_0 + \frac{1}{2} \\ \qquad\qquad\qquad\qquad\qquad\qquad + u_{\alpha/2}\sqrt{tp_0 q_0}. \end{cases}
$$

When n is sufficiently large, we can calculate the Pearson statistic of goodness of fit to a theoretical binomial of parameter p by [5].

$$
\begin{aligned}
T(t) &= \frac{(x - tq_0)^2}{tq_0} + \frac{(y - tp_0)^2}{tp_0} \\
&= \frac{(y - tp_0)^2}{tp_0 q_0} ,
\end{aligned}
$$

$T(t)$ is the square of the standardized variable $y - tp_0/\sqrt{tp_0 q_0}$ which is asymptotically normal $N(0.1)$. $T(t)$ is a χ^2 with one degree of freedom. The critical function can be defined by:

$$
\varphi(y,t) = \begin{cases} 1 \text{ when } T(t) > \chi_\alpha^2, \\ 0 \text{ when } T(t) < \chi_\alpha^2, \end{cases}
$$

χ_α^2 is the value of the $1 - \alpha$ percentile for a χ^2 with one degree of freedom.

We can remark that the chi-square test is automatically a two-sided hypothesis test because by squaring the quantity $y - tp_0 \times (\sqrt{tp_0 q_0})^{-1}$ the sign deviation is lost.

Example 3.2.2

The survey on roads with and without plantations gives

$$x = 1,487, \quad y = 1,142, \quad \text{and} \quad p_0 = 0.44.$$

By using the normal approximation, we calculate the two cut-offs $C_1(t)$ and $C_2(t)$:

$$C_1(t) = 2,629 \times 0.44 - 1.96\sqrt{2,629 \times 0.44 \times 0.56} = 1,106.87,$$

$$C_2(t) = 2,629 \times 0.44 + 1.96\sqrt{2,629 \times 0.44 \times 0.56} = 1,206.6.$$

The critical function can be written as

$$\varphi(y,t) = \begin{cases} 1 \text{ when } y < 1,107 \text{ or } y > 1,207, \\ \\ 0 \text{ when } 1,107 < y < 1,207, \end{cases}$$

with $y = 1,142$, we accept the null hypothesis and conclude a non-significant difference between the number of accidents on roads with and without plantations.

The Pearson statistic has for value

$$T(t) = \frac{(1,142 - 2,629 \times 0.44)^2}{2,629 \times 0.44 \times 0.56} \quad 0.336.$$

$T(t)$ is far below any significant level, so we can accept the null hypothesis.

4. COMPARISON OF FOUR POISSON DISTRIBUTIONS

Consider four independent Poisson variables X_1, Y_1, X_2, Y_2 with parameters $\lambda_1, \mu_1, \lambda_2, \mu_2$ and realisations x_1, y_1, x_2, y_2 which can be presented in a 2×2 table:

	before	after
Control	x_1	y_1
Experimental	x_2	y_2

$t_1 = x_1 + y_1,$

$t_2 = x_2 + y_2,$

$t = y_1 + y_2.$

Two composite hypotheses can be considered:

$$H_1: \frac{\mu_1}{\lambda_1} = \frac{\mu_2}{\lambda_2} , \qquad\qquad K_1: \frac{\mu_2}{\lambda_2} < \frac{\mu_1}{\lambda_1} ;$$

$$H_2: \frac{\mu_1}{\lambda_1} = \frac{\mu_2}{\lambda_2} , \qquad\qquad K_2: \frac{\mu_2}{\lambda_2} \neq \frac{\mu_1}{\lambda_1} .$$

The null hypothesis expresses that the evolutions between the two periods before and after are the same on the experimental and the control. The alternatives express either a smaller or a different evolution between the two groups.

The joint distribution of the four Poisson variables belongs to a multiparameter exponential form.

To find an unbiased UMP test for these hypotheses we can use a conditional test by fixing both the sums $X_1 + Y_1 = t_1$, $X_2 + Y_2 = t_2$ on the control and on the experimental. As the distributions are independent, the conditional joint distribution of Y_1 and Y_2, given $X_1 + Y_1 = t_1$ and $X_2 + Y_2 = t_2$, is equal to the product of two binomial distributions:

$$P(Y_1 = y_1, Y_2 = y_2, X_1 + Y_1 = t_1, X_2 + Y_2 = t_2)$$

$$= \binom{t_1}{y_1}\left(\frac{\mu_1}{\lambda_1 + \mu_1}\right)^{y_1}\left(\frac{\lambda_1}{\lambda_1 + \mu_1}\right)^{t_1 - y_1}\binom{t_2}{y_2}\left(\frac{\mu_2}{\lambda_2 + \mu_2}\right)^{y_2}$$

$$\times \left(\frac{\lambda_2}{\lambda_2 + \mu_2}\right)^{t_2 - y_2}$$

$$= \binom{t_1}{y_1}\binom{t_2}{y_2}\left(\frac{\lambda_1}{\lambda_1 + \mu_1}\right)^{t_1}\left(\frac{\lambda_2}{\lambda_2 + \mu_2}\right)^{t_2}$$

$$\times \exp\left[y_2\left(\log\frac{\mu_2}{\lambda_2} - \log\frac{\mu_1}{\lambda_1}\right) + (y_1 + y_2)\log\frac{\mu_1}{\lambda_1}\right] .$$

The problem is transformed into a testing problem of homogeneity of binomial distributions. The two hypotheses and alternatives $(H_1, K_1), (H_2, K_2)$ can be tested concerning the parameter $\vartheta = \log((\mu_2/\lambda_2)/(\mu_1/\lambda_1))$ or equivalently concerning the ratio $\rho = (\mu_2/\lambda_2)/(\mu_1/\lambda_1)$. The null hypothesis is $\vartheta = 0$ or $\rho = 1$. Putting $U = Y_2$ and $T = Y_1 + Y_2$, the test is carried out in terms of the conditional distribution of Y_2, given $Y_1 + Y_2 = t$:

$$P(Y_2 = y_2, X_1 + Y_1 = t_1, X_2 + Y_2 = t_2, Y_1 + Y_2 = t)$$

$$= \binom{t_1}{t - y_2}\binom{t_2}{y_2}\binom{t_1 + t_2}{t}, \qquad y_2 = 0, 1, \ldots, t.$$

It is the hypergeometric distribution of Y_2 given three marginal sums t_1, t_2, t in a 2×2 table. It is the form of the exact test proposed by Fisher for 2×2 tables. The important result is that the same test can be used in different situations: all margins fixed, two margins fixed (homogeneity), or no margins fixed (double dichotomy) by application of conditional test results [6].

4.1 One-Sided Hypothesis

$$H_1: \frac{\mu_1}{\lambda_1} = \frac{\mu_2}{\lambda_2} \, , \qquad\qquad K_1: \frac{\mu_2}{\lambda_2} < \frac{\mu_1}{\lambda_1} \, .$$

4.1.1 Small Samples

The critical function of an unbiased UMP test for testing

$$H_1: \frac{\mu_1}{\lambda_1} = \frac{\mu_2}{\lambda_2} \qquad \text{against} \qquad K_1: \frac{\mu_2}{\lambda_2} < \frac{\mu_1}{\lambda_1}$$

is determined by

$$\varphi(y, t, t_1, t_2) = \begin{cases} 1 & \text{when } y < C_0(t, t_1, t_2) \\ \gamma_0(t) & \text{when } y = C_0(t, t_1, t_2), \\ 0 & \text{when } y > C_0(t, t_1, t_2), \end{cases}$$

with C_0 and γ_0 determined by the formula

$$\sum_{y=0}^{C_0-1} \frac{\binom{t_1}{t-y}\binom{t_2}{y}}{\binom{t_1+t_2}{t}} + \gamma_0 \frac{\binom{t_1}{t-C_0}\binom{t_2}{C_0}}{\binom{t_1+t_2}{t}} = \alpha.$$

The selection of C_0 and γ_0 for a critical level α follows the same rule as for the binomial case (cf. 3.1.1): choose C_0 for which the cumulative probability is more than α and adjust the test size to α with an auxiliary randomization of probability γ_0. Tables have been computed for small samples. Finney [4] gives the value of y required to reject the null hypothesis for value of t, t_1, t_2, up to 15 for a single-tailed test of size $\alpha \leqslant 0.05$, 0.01 and in addition with the exact size in each case. Armsen [1] gives the same information for one or two-tailed tests of size 0.05 or 0.01 with t_1, t_2, t, ranging to 50.

Example 4.1.1

For a study about the effectiveness of guard rails on motorways, we observed the number of accidents with fatality presented in a 2×2 table:

	Before	After	
Control	10	8	$18 = t_1$
Experimental	7	1	$8 = t_2$

$$9 = t.$$

We calculate the different probabilities of Y following an hypergeometric distribution with the formula:

$$P(Y = y) = \frac{\binom{18}{9 - y}\binom{8}{y}}{\binom{26}{9}} = \frac{18!9!8!17!}{26!(9 - y)!y!(9 + y)!(8 - y)!} ,$$

which leads to the results:

y	$P(Y = y)$	$P(Y \leqslant y)$
0	0.016	0.016
1	0.112	0.128

C_0 is equal to 1 with $\gamma_0 = \dfrac{0.05 \quad 0.016}{0.112} = 0.303$.

The critical function can be written as

$$\varphi(y,9,18,6) = \begin{cases} 1 & \text{if } y < 1, \\ 0 & \text{if } y > 1, \\ 0.303 & \text{if } y = 1. \end{cases}$$

With $\alpha = 0.05$, $y = 1$, so we have to make a random experiment with probability 0.303 of rejection. We have more chance to accept the null hypothesis.

4.1.2 Large Samples

When $t_1 + t_2 = n$ is sufficiently large, Y_2 is asymptotically normal with mean tt_2/n and variance $tt_1t_2(n - t)/n^2(n - 1)$. By using the standardized variable

$$v = \left(y_2 - \frac{tt_2}{n}\right)\left(\frac{tt_1t_2(n - t)}{n^2(n - 1)}\right)^{-\frac{1}{2}}.$$

we can define a critical function for an approximate UMP unbiased
test by

$$\varphi(y,t,t_1,t_2) = \begin{cases} 1 \text{ when } y < \dfrac{tt_2}{n} - u_\alpha \left(\dfrac{tt_1t_2(n - t)}{n^2(n - 1)}\right)^{\frac{1}{2}}, \\ 0 \text{ when } y > \dfrac{tt_2}{n} - u_\alpha \left(\dfrac{tt_1t_2(n - t)}{n^2(n - 1)}\right)^{\frac{1}{2}}. \end{cases}$$

We can introduce a continuity correction by subtracting 1/2 from
the expected mean tt_2/n.

Example 4.1.2

One year of observation before and after concrete grooving
on control and experimental sections of motorways leads to the
following 2×2 table of number of accidents:

	Before	After	
Control	71	77	148
Experimental	56	34	90
	111	238	

The expected mean is equal to $\dfrac{tt_2}{n} = \dfrac{111 \times 90}{238} = 41.97$, and the
variance is equal to $\dfrac{tt_1t_2(n - t)}{n^2(n - 1)} = \dfrac{111 \times 148 \times 90 \times 127}{(238)^2 \times 237} = 13.98 = (3.74)^2$. For a critical level $\alpha = 0.05$, $u_\alpha = 1.65$ and

$$C_0 = 41.97 - 1.65 \times 3.74 = 37.8.$$

An approximate unbiased UMP test is defined by the critical function

$$\varphi(y,111,148,90) = \begin{cases} 1 \text{ if } y < 37.8, \\ 0 \text{ if } y > 37.8. \end{cases}$$

As $y = 34$, we reject the null hypothesis with a critical level
$\alpha = 0.05$ and conclude a significant reduction of accidents due to
concrete grooving of pavements of motorways.

4.2 Two-Sided Hypothesis

$$H_2: \frac{\mu_1}{\lambda_1} = \frac{\mu_2}{\lambda_2} , \qquad\qquad K_2: \frac{\mu_1}{\lambda_1} \neq \frac{\mu_2}{\lambda_2} .$$

4.2.1 Small Samples

Adopting the formula of 3.2.1 with the hypergeometric distribution instead of the binomial distribution, we can define critical functions φ which give unbiased UMP tests.

4.2.2 Large Sample Chi-Square Test

For large samples, we use the chi-square test of homogeneity between two binomial laws [5,7].

	Before	After	
Control	x_1	y_1	t_1
Experimental	x_2	y_2	t_2
			n

The null hypothesis is $H_2: p_c = p_e$ against $K_2: p_c \neq p_e$ between two binomial laws $B(p_c, t_1)$ on the control, $B(p_e, t_2)$ on the experimental. We need an estimate of the common value of p_e and p_c for the null hypothesis:

— on the basis of both control and experimental samples by

$$p^* = \frac{y_1 + y_2}{n} ;$$

— on the basis of the control sample by

$$p' = \frac{y_1}{t_1} .$$

The Pearson statistic for an homogeneity chi-square test is

$$\chi^2_{t_1, t_2} = \frac{(x_1 - t_1 q^*)^2}{t_1 q^*} + \frac{(y_1 - t_1 p^*)}{t_1 p^*} + \frac{(x_2 - t_2 q^*)^2}{t_2 q^*}$$

$$+ \frac{(y_2 - t_2 p^*)^2}{t_2 p^*} = \qquad\qquad \text{(Contd)}$$

(Contd) $\quad = \dfrac{(y_1 - t_1 p^*)^2}{t_1 p^* q^*} + \dfrac{(y_2 - t_2 p^*)^2}{t_2 p^* q^*}$

$\quad\quad = \dfrac{(t_1 + t_2)}{t_1 t_2} \dfrac{(x_1 y_2 - x_2 y_1)^2}{(x_1 + x_2)(y_1 + y_2)}$,

with q^* estimated and $\chi^2_{t_1, t_2}$ calculated on both experimental and control.

$\chi^2_{t_1, t_2}$ is asymptotically a χ^2 with one degree of freedom. The critical function defining an asymptotically UMP test is determined by

$$\varphi(y, t_1, t_2) = \begin{cases} 1 \text{ when } \chi^2_{t_1, t_2} > \chi^2_\alpha, \\[2mm] 0 \text{ when } \chi^2_{t_1, t_2} < \chi^2_\alpha. \end{cases}$$

The Tanner chi-square test is defined by the statistic [9]

$$c^2_{t_1, t_2} = \dfrac{(x_2 - t_2 q')^2}{t_2 q'} + \dfrac{(y_2 - t_2 p')^2}{t_2 p'}$$

$$= \dfrac{(y_2 - t_2 p')^2}{t_2 p' q'} ,$$

with q' estimated on the control and $c^2_{t_1, t_2}$ calculated on the experimental.

When $t_1 \rightarrow \infty, t_2 \rightarrow \infty$, the asymptotical distribution $C^2(t_1, t_2)$ [2] is no more a χ^2 but $(1 + \tau)\chi^2$ with $\tau = \lim\limits_{\substack{t_1 \rightarrow \infty \\ t_2 \rightarrow \infty}} t_1/t_2$.

The limit distribution depends on the ratio of the experimental to the control sizes. This statistic must be used only when the control size is sufficiently large according to the experimental size.

Example 4.2.2

Taking again the 2 × 2 table for the concrete grooving experiment:

	Before	After	
Control	71	77	148
Experimental	56	34	90
	127	111	238

The Pearson statistic is equal to:

$$\chi^2_{148,90} = \frac{(148 + 90)}{148 \times 90} \frac{(71 \times 34 - 56 \times 77)^2}{127 \times 111} = 4.566$$

For a critical level $\alpha = 0.05$, 4.56 is higher than the critical value $\chi^2_\alpha = 3.84$. We reject the null hypothesis and conclude a significant reduction of accident due to grooving. The Tanner chi-square test is equal to:

$$C^2_{148,90} = \frac{(34 - 90 \times 0.52)^2}{90 \times 0.52 \times 0.48} = 7.32.$$

$C^2_{148,90}$ must be compared with $(1 + \tau)\chi^2_\alpha$ with $\tau = \frac{90}{148} = 0.61$, that is to say with 6.18 instead of 3.84.
    We reject the null hypothesis also.

4.2.3 Maximum Likelihood Ratio Test

    Let us start again with the following 2×2 table:

	Before	After
Control	x_1	y_1
Experimental	x_2	y_2

where x_1, y_1, x_2, y_2 are the realisations of 4 independent Poisson variables of parameters $\lambda_1, a\lambda_1, \mu_1, b\mu_1$. The null hypothesis is that the evolution is the same on the control and the experimental: H_2: $a = b$ against K_2: $a \neq b$ [7].
    To test the hypothesis, we take the ratio of the maximum of the likelihood in two cases H and K expressed by $\Lambda = \mathrm{Max}L/\mathrm{Max}L$
$$\qquad\qquad\qquad\qquad\qquad\qquad\qquad\qquad H \qquad K$$
with

$$\log L = - \lambda_1 + x_1\log\lambda_1 - \mu_1 + y_1\log\mu_1 - a\lambda_1 + y_1\log a\lambda_1$$

$$\cdot \quad - b\mu_1 + y_2\log b\mu_1 - \log(x_1!x_2!y_1!y_2!).$$

By differentiating the logarithm of the likelihood and setting to zero the different derivatives, we obtain estimates of the parameters according to the hypothesis:

(1) K_2: $a \neq b$

$$\hat{\lambda}_1 = x_1, \qquad \widehat{a\lambda_1} = y_1, \qquad \hat{\mu}_1 = x_2, \qquad \widehat{b\mu_1} = y_2;$$

(2) H_2: $a = b$

$$\hat{a} = \hat{b} = \frac{y_1 + y_2}{x_1 + x_2}, \qquad \hat{\lambda}_1 = \frac{(x_2 + y_2)(x_1 + x_2)}{x_1 + x_2 + y_1 + y_2},$$

$$\hat{\mu}_1 = \frac{(x_1 + y_1)(x_1 + x_2)}{x_1 + x_2 + y_1 + y_2}.$$

The likelihood ratio can be written

$$\Lambda = \frac{e^{(x_2+y_2)\log(x_2+y_2)+(x_1+x_2)\log(x_1+x_2)+(y_1+y_2)\log(y_1+y_2)}}{e^{x_1\log x_1+x_2\log x_2+y_1\log y_1+y_2\log y_2+(x_1+x_2+y_1+y_2)\log(x_1+x_2+y_1+y_2)}}.$$

The quantity $-2\log\Lambda$ which is asymptotically distributed as a χ^2 with one degree of freedom allows us to construct a critical function for an asymptotical test:

$$\varphi(x_1,x_2,y_1,y_2) = \begin{cases} 1 \text{ if } -2\log\Lambda > \chi_\alpha^2, \\ 0 \text{ if } -2\log\Lambda < \chi_\alpha^2. \end{cases}$$

5. CONCLUSION

The comparison of two or four Poisson variables which arises in effectiveness evaluation after designing an experiment can be resolved by the conditional test theory.

For small sampels and for one- or two-sided hypotheses, UMP unbiased randomized tests can be constructed from binomial or hypergeometric distributions with the aid of statistical tables.

For large samples and for one- or two-sided hypotheses, the normal approximation of binomial or hypergeometric distributions provides asymptotical tests.

For large samples, the classical chi-square tests and the

maximum likelihood ratio tests can be easily constructed for a two-sided hypothesis.

BIBLIOGRAPHY

[1]   ARMSEN, P.   "Tables for significance tests of 2 × 2 contin-
            gency tables" in Biometrika, 1955, Vol. 42,
            pp. 494-511.

[2]   CHASE   "On the chi-square test when the parameters
            are estimated independently of the sample", in
            J.A.S.A., 1972, Vol. 67, pp. 609-611.

[3]   CONOVER, W.J. and rejoinders: "Some reaons for not using
            the Yates continuity correction on 2 × 2 contin-
            gency tables", in J.A.S.A., 1974, Vol. 69,
            pp. 374-382.

[4]   FINNEY, D.J. "The Fisher-Yates test of significance in
            2 × 2 contingency tables" in Biometrika, 1948,
            Vol. 35, p. 145.

[5]   GARWOOD, F. and NEWBY, R.F. "Utilisation du χ^2 pour la
            comparaison des fréquences d'accidents" in
            Symposium OCDE sur l'utilisation des méthodes
            statistiques dans l'analyse des accidents de la
            route, OCDE, Paris, 1970, pp. 37-44.

[6]   KENDALL, M.G. and STUART, A. "The Advanced Theory of
            Statistics", Vol. 2, Chapter 33, C. Griffin
            and Co., Ltd.,

[7]   LASARRE, S. "A propos des tests statistiques sur variables
            poissoniennes, utilisés dans le domaine de la
            Sécurité Routière" in Revue de Statistique
            Appliquée, 1977, Vol. XXV, n° 3, pp. 55-74.

[8]   LEHMANN, E.L. "Testing statistical Hypotheses", John
            Wiley and Sons, Inc..

[9]   TANNER, J.C. "A problem in the combination of accident
            frequencies" in Biometrika, 45, 1958,
            pp. 331-342.

[10] ULMO, J. and BERNIER, J. "Eléments de décision statis-
            tique", P.U.F..

A DATA REDUCTION METHOD BASED ON INFORMATION THEORY

T. Bui Quoc

Organisme National de Sécurité Routière, BP.28
94114 - Arcueil Cedex

ABSTRACT. 1/Introduction and overall considerations. 2/Principle
of the data reduction method. 3/Applications of the method on data
provided by two surveys. 4/Discussion.

1. INTRODUCTION

During these recent years, there is a tendency to include more and
more data, resulted either from sample surveys or from exhaustive
inquiries, in socio-economical analysis. This trend results essentially from the concern of keeping collective consensus or preferences
according to several criteria, thus leading to use the techniques
of multidimensional analysis more frequently.

By the introduction and development of these techniques which become
more and more elaborated, the corresponding data files become more
and more voluminous. Furthermore, using too many variables in multidimensional analysis is sometimes impossible due to technical
limitations - input specifications of computer programs, capacity
of computer memories - and very frequently involve quite annoying
consequences : too high cost of computing time, difficulties in
dealing with results,... Actually a certain number of variables,
considered separately or in combination, could be ignored without
modifying the concepts expected to come out : voluminous data files
always contain redundances between variables.

This problem of searching correlation or redundance between variables is known since long time ago and many measurements of dependence between variables are available such as Spearman-Kendall's
correlation coefficient, Pearson's chi-square, Tschuprow's T
coefficient,... On the other hand, some methods of data analysis

allow to classify variables into homogenous clusters such as principal components analysis, correspondence analysis, classification,...

We propose to use a data reduction method less sophisticated which is nevertheless based on rigourous criteria provided by Information Theory.

2. PRINCIPLE

Let Ω be a set of events provided with a measure of probability p and let X and Y be two partitions of Ω :

$$X = \{X_i, \ i = 1,\ldots I \mid X_i \cap X_{i'} = \emptyset, \forall_i \neq_{i'}, \text{ and } U_{i \in I} \ X_i = \Omega \}$$

$$Y = \{Y_j, \ j = 1,\ldots J \mid Y_j \cap Y_{j'} = \emptyset, \forall_j \neq_{j'}, \text{ and } U_{j \in J} \ Y_j = \Omega \}$$

We note H(X) the Shannon's entropy of the random variable X :

$$H(X) = - \Sigma_{i=1}^{I} \ p(X_i) \ \text{Log} \ p(X_i) \tag{1}$$

The entropy H(X) is in some ways the measure of the average "uncertainty" of the random variable X : if X is perfectly determined, H(X) will be equal to nil and if X is completely undetermined, i.e. each event X_i having the same probability, H(X) is maximum and equal to log (I).

We note H(Y|X) the conditionnal entropy of Y for X known :

$$H(Y|X) = - \Sigma_{i=1}^{I} \ p(X_i) \ \Sigma_{j=1}^{J} \ p(Y_j|X_i) \ \text{Log} \ p(Y_j|X_i) \tag{2}$$

If X and Y are independent, knowing X does not lead to any information on Y and we will get H(Y|X) equal to H(Y). On the contrary, if knowing X determines perfectly Y, H(Y|X) will be nil.

The mutual information between random variables X and Y can be written as following :

$$MI(X,Y) = H(X) - H(X|Y) = H(Y) - H(Y|X) \tag{3}$$

Thus the more correlated are two variables, greater their mutual information will be.

$$\text{The ratio} \quad RI(X/Y) = \frac{MI(X,Y)}{H(Y)} = \frac{H(Y) - H(Y|X)}{H(Y)} \tag{4}$$

could be considered as a relative information earnings furnished by random variable X over random variable Y.

The principle of the data reduction method consists of selecting random variables which will provide highest relative information about all the other variables of the file.

In the case of N random variables, the computational procedure is the following :

a) Building the matrix of the N x N ratio RI between all the couples of variables.

b) For each variable, calculate the sum of relative information.

c) Select the variable having the highest sum of relative information and calculate the cumulative percentage of information obtained from this variable. Cancel the row and the column corresponding to this variable in the matrix and reiterate to step b) of the procedure for the remaining variables.

The cumulative percentage of information CPI_k obtained after selecting k first variables is given by :

$$CPI_k = CPI_{k-1} + (1 - CPI_{k-1}) \frac{SRI_k}{N-k+1}$$

where

CPI_{k-1} = cumulative percentage of information obtained at $(k-1)$th step

SRI_k = sum of relative information of the variable selected at kth step

CPI_o = 0

3. APPLICATIONS

This data reduction method was used in a 1978 survey conducted by a team of ONSER's sociologists which is considered as a preliminary step to the setting up of an instrument in order to measure public reception of spots on Television about road safety problems. For each of the six TV spots about seat belt, fifty persons filled a form of 58 items, each item being a couple of opposite adjectives inserted in a seven point scale, e.g. interesting vs not interesting. clear vs confused, simple vs complicated... The data reduction method finds here an application field extremely favourable for two reasons
- a small size sample serving as a preliminary step to further full size surveys
- visible resemblances between items of the form : many items have,

knowingly or not, great similitudes. For example, we can cite the 22nd item (likely vs unlikely) and the 36th item (incredible vs likely), or the 43rd item (tragical vs comical) and the 57th item (burlesque vs serious), ...

Results obtained from the reduction method reveal the velocity of the computational procedure : more than 95 % of information are provided by no more than 10 variables and more than 99 % of information are given by no more than 15 variables among the initial 58 items of the form. It denotes that a great number of items could eventually be omitted in further surveys without modifying information collected. On the other hand, a simple analysis of the six series of selected variables shows a dichotomy of the six TV spots into two separate clusters, the one being constitued of spots containing tragical and violent images such as crashed cars or injured passengers which cannot be found in the other. Moreover, this dichotomy is confirmed by the results of a correspondence analysis in progress.

According to these results, one can proceed to reduce the number of questions in report form of further surveys or to simplify further multidimensional analysis for we have to keep in mind that reduction method is not in itself a data analysis technique.

In order to validate it we made a second application in the following manner :

- apply the method on variables provided by another survey which already have results given by a principal component analysis

- execute a second principal component analysis on a reduced number of variables which are selected by the method

- compare the results of the two analysis in order to verify their steadiness

The survey in question was conducted in 1977 on a sample of 354 drivers to determine their opinion about automobile, speed limit, repression and causes of road accidents. The first principal component analysis opposes controversial behaviour to non controversial attitudes on the first factorial axe and reveals, on the first factorial plan, the existence of two secondary axes, the one contrasting a safe behaviour to a lightly risky one, the other opposing road network responsibility to driver's training.

The reduction method, when applied on 81 variables previously mentionned, shows that 90 % of the information are given by only 40 variables. The computational procedure is obviously slower than for the first application because various subjects were treated in this second survey. Admitted the loss of 10 % of the information, the principal component analysis of the 40 selected variables reveals a noticeable steadiness of the results as shown in the first factorial plan (figure 1) : the respective positions of variables are quite the same in the two analysis (direction of arrows in figure 1) and on the first factorial axe, controversial behaviour is opposed to non controversial one. In upper left quarter of the figure 1 are gathered variables associated to a safe behaviour and in lower left quarter are found variables implicating road responsability in accident genesis ; by deduction, the lower right quarter denotes a lightly risky behaviour and the upper right quarter implicates driver's training.

72

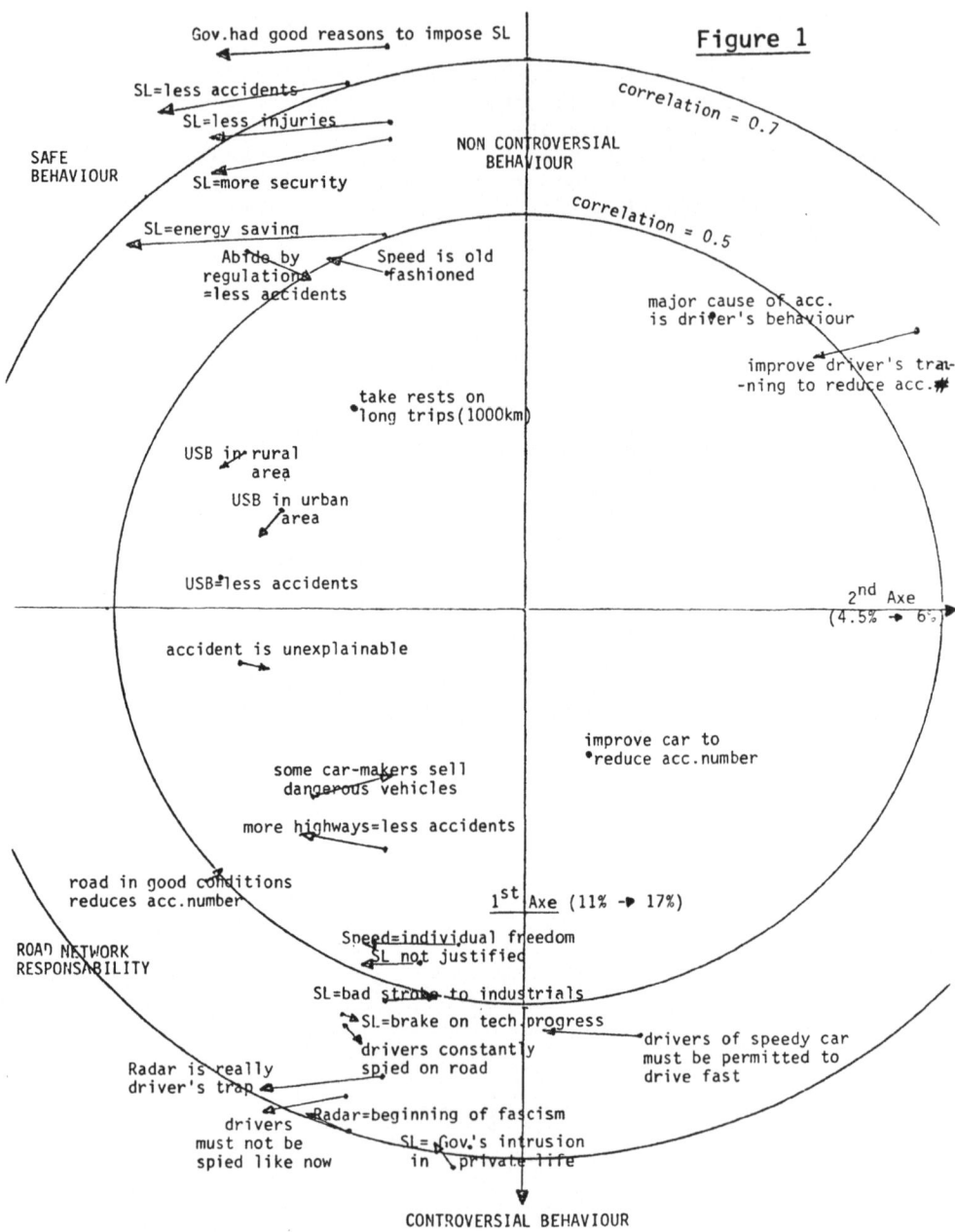

Figure 1

SL = Speed Limit
USB = Use Seat Belt

4. <u>DISCUSSION</u>

The data reduction method based on Information Theory seems to be
efficient and reliable. It could be used as a preliminary step to
improve efficiency of surveys by reducing the number of items in
report form, or to simplify utilization of sophisticated multidimensional analysis by selecting the most interesting variables to
be studied without modifying the expected results. Nevertheless,
in the first case one must be certain that the studied sub-sample
is absolutely representative and in the second case, one must avoid
to apply the method on heterogeneous variables.

It seems to be possible, and even desirable, to improve the method
by using other criteria to select variables or items rather than
the sum of relative information.

The actual computer program, drafted by J.P. Aimetti and J.C.Brousse
is part of a large package named N.L.T. (New Language for Tabulation)
which major aim is cross tabulation of data, and consequently
cannot be used without appropriate environment. However the principle of this data reduction method raises no major difficulty to the
realization of a computer program.

MEASURING EXPOSURE

Diccon Bancroft
Consumers Union of U.S., Inc.

Johan Riemersma
Institute for Perception TNO,
Soesterberg, The Netherlands

INTRODUCTION

The classical contingency table is a multidimensional array
of counts in which the cells consist of all combinations at all
levels of some cross-classifying variables. In two dimensions
this reduces to a rectangular array of counts. But not all tables
come in this form, so to examine the flexibility of the MDI approach, an unconventional table was analysed.
The data set chosen is a triangular array of counts, a (partial) table from the work of Koornstra (1973a,b) on modelling interactions between exposure and proneness in accident data. This
array has already been analysed using different techniques (Koornstra 1973b) and it is of some interest to contrast this earlier
analysis with the present one.

EXPOSURE AND PRONENESS

One of the major problems in accident investigations is to
gather control data, with which the occurrence of other characteristics of accidents must be related before useful conclusions can
be reached. These control data take very different forms, depending on the particular investigation, for example, general trends
in accident occurrence, vehicle miles of travel, traffic densities,
accident frequencies at control sites, and so on. They can be
thought of as covariates that the investigator wants to partial
out from the effect under investigation. In investigations that
attempt to establish different accident probabilities for (classes
of) individuals, the covariate is called 'exposure'. Essentially,

it is a measure of the opportunity of being involved in an accident, while proneness is the proportion of occasions one was in fact involved in an accident. Usually, at best, rough estimates of exposure are available. Vehicle miles of travel, for instance, is much too crude a measure of exposure when one is interested in rear-end collisions at night, but more precise exposure measures are not available or can only be obtained at great cost. In the past, to avoid these difficulties, several attempts have been made to generate estimates of exposure from the accident data, by making assumptions about the ratio of the number of single-vehicle accidents to the number of two-vehilce accidents, and/or the ratio of guilty to innocent drivers for several classes of drivers. The work of Koornstra (1973a,b) is along these lines.

THE DATA

The table consists of counts of fatal two-vehicle accidents involving only passenger cars in the Netherlands for the years 1968-1970. The cells of the table correspond to age-sex combinations of the two drivers, and the table can be thought of as a folded or collapsed version of an unobserved full table in which only one of each pair of drivers is distinguished in some manner (for example, was found guilty, or suffered the most damage, etc.). This underlying table is square, and the observed triangular table results from summing entries symmetrical about the diagonal. The folded, triangular table is displayed in Table 1.

TABLE 1

*Two-Car Fatal Accidents in the Netherlands
1968-1970 Classified by Age and Sex of Drivers*

Class sex	age	Code	m_{21}	m_{30}	m_{40}	m_{50}	m_{60}	m_{70}	w_{23}	w_{37}	w_{55}
m	18-24	m_{21}	52								
m	25-34	m_{30}	83	41							
m	35-44	m_{40}	71	46	20						
m	45-54	m_{50}	42	60	37	17					
m	55-64	m_{60}	31	36	22	19	7				
m	>64	m_{70}	21	22	18	13	3	1			
w	18-29	w_{23}	10	11	12	3	5	1	2		
w	30-44	w_{37}	10	11	5	9	6	4	2	0	
w	>44	w_{55}	8	13	9	11	4	4	1	0	2

KOORNSTRA'S APPROACH

Let e_i be a measure of exposure for driver class i, and let p_i be the measure of exposure-corrected accident involvement (accident proneness). The model that Koornstra proposes is that the underlying table has entries with expected value

$$t_{ij} = (p_i e_i)e_j \tag{1}$$

so that the symmetrised table, before folding, has entries with expected value

$$t_{ij} = (t_{ij} + t_{ji})/2 = (p_i + p_j)e_i e_j/2. \tag{2}$$

After folding the expected values are

$$x_{ii} = p_i e_i e_i \tag{3.1}$$

down the diagonal, and

$$x_{ij} = (p_i + p_j)e_i e_j \tag{3.2}$$

in the lower triangle.

He fits this model by an iterative least squares method, trying to minimze Neyman's modified chi-squared statistic for the observed and estimated x_{ij}'s, and obtains a good fit to the data (Table 4, p. 177, Koornstra 1973b). The values he reports are reproduced in Table 2.

TABLE 2
Koornstra's Parameter Estimates

Code	m_{21}	m_{30}	m_{40}	m_{50}	m_{60}	m_{70}	w_{23}	w_{37}	w_{55}
e_i	3.123	8.697	6.615	2.167	2.082	1.357	0.479	0.996	0.741
p_i	2.857	0.314	0.339	2.647	1.470	1.525	2.702	0.757	1.604

$$X_m^2 = 45.7 \qquad d_f = 28$$

THE MDI APPROACH

It is not possible to write Koornstra's model directly as a log-linear model because of the non-additivity in the logarithms of the expected cell counts induced by the folding of the unobserved underlying table. But a variety of log-linear models can be considered once the proneness parameters p_i are fixed. In the following, the p_i's are set to the value 1 in equations (1-3), corresponding to a model in which all driver classes have the same proneness. These models are then a restricted version of that of Koornstra.

The MDI approach involves specifying a constraint matrix C and a reference distribution π, such that $\ln(p/\pi)$ is linear in the parameters. The usual reference distribution is the Uniform, and from equations (3) for this triangular table should be proportional to the distribution with ones down the diagonal and twos off the diagonal, corresponding to the Uniform distribution in the unobserved full table.

For the triangular table the constraints of the constraint matrix can be thought of as fixing counts of accidents or drivers. The cell indexing and a variety of possible rows for the constraint matrix C are given in Table 3. What the constraints fix is listed below:

L: the total number of accidents (half the number of drivers)
ϑ_1: the number of m_{20}'s
ϑ_2: the number of m_{30}'s
ϑ_3: the number of m_{40}'s
ϑ_4: the number of m_{50}'s
ϑ_5: the number of m_{60}'s
ϑ_6: the number of m_{70}'s
ϑ_7: the number of w_{23}'s
ϑ_8: the number of w_{37}'s
ϑ_9: the number of male drivers
ϑ_{10}: the number of accidents involving only male drivers.

The constraints used specify the model that is fit. For constraints involving only counts of drivers, driver 1 — driver 2 characteristics are modelled as being stochastically independent within accident type. Inclusion of constraints involving accident counts (other than L) specify accident types. For example, ϑ_{10} separates out accidents between male drivers only, and together with the total count of males (from ϑ_9 or $\vartheta_1,\ldots,\vartheta_6$) allows different accident probabilities for male-male, male-female and female-female accidents. Thus constraints fixing the counts of accidents in various age-sex classes can be used to some extent in modelling proneness. The major model of interest uses constraints L and $\vartheta_1 - \vartheta_6$. With these constraints accidents are modelled as occurring at random between drivers from the observed population of drivers involved in two-car fatal accidents.

Parameter estimates for these constraints using the folded uniform reference distribution were produced using the program KULLITR. Because it was not known what the effect of zero cell frequencies might be, the original triangular table was analysed, as was a smoother table produced by adding 1 to all 45 original cell counts. For both these tables a good fit was obtained, and examination of the deviations from the fit showed no abnormally large values and no clear patterns. The results of the analysis are summarised in Table 4.

TABLE 3

Cell Indexing and Constraint Matrix for MDI Approach

ij indexing

code	m21	m30	m40	m50	m60	m70	w23	w37	w55
m21	11								
m30	21	22							
m40	31	32	33						
m50	41	42	43	44					
m60	51	52	53	54	55				
m70	61	62	63	64	65	66			
w23	71	72	73	74	75	76	77		
w37	81	82	83	84	85	86	87	83	
w55	91	92	93	94	95	96	97	98	99

C matrix

TABLE 4

MDI FITS

Code:		m_{21}	m_{30}	m_{40}	m_{50}	m_{60}	m_{70}	w_{23}	w_{37}	w_{55}
Parameter:	L	ϑ_1	ϑ_2	ϑ_3	ϑ_4	ϑ_5	ϑ_6	ϑ_7	ϑ_8	ϑ_9
Table										
Original	-0.099	1.951	1.908	1.572	1.440	0.953	0.488	-0.097	-0.139	-
Smoothed	0.186	1.807	1.765	1.440	1.313	0.852	0.426	-0.081	-0.116	-

Goodness of Fit	X^2	*MDI Statistic*
Original Table	34.6330	37.6462
Smoothed Table	34.2773	33.6004

on 36 degrees of freedom

REANALYSIS OF KOORNSTRA'S MODEL

Koornstra's model as specified in equation (3) is not log-linear. However, the underlying model as specified in equation (1) is. The full table is not observed, so model fitting via the EM algorithm (Dempster, Laird and Rubin, 1977) might be appropriate. In the full table the M step corresponds to fitting the usual independence model for two-way tables. The E step then adjusts the fitted values in the full table so that the corresponding folded table is the observed one. The convergence of the algorithm can be accelerated by Newton-Raphson type procedures.

The resulting estimates are maximum-likelihood under the usual multinomial assumptions, and as such lead to Information Statistics directly comparable to those obtained for the restricted model via the MDI approach. The results obtained are summarised in Table 5.

The major difference between the fits to the original table and to the smoothed table occur for the m_{70} and the w_{37} groups. For the m_{70} group the proneness estimate from the smoothed table is half that from the original table, and for the w_{37} group in the original table the best fit is obtained with a proneness of zero. The parameters for the other groups seem very similar between the two tables. There were no clear patterns in the deviations from the fits and no individual cell deviation was abnormally large. The low value of the goodness of fit statistics with 28 degrees of freedom and 17 parameters fitted is, perhaps, an indication of overfitting.

DISCUSSION

To ensure comparability in the method of fitting and in the goodness of fit statistics, the values obtained in the reanalysis of Koornstra's model using the EM algorithm will be used. The goodness of fit statistics both for the original table and for the table smoothed by adding 1 to all the cell counts are collected in Table 6. The chi-squared values are close to the MDI statistics, as expected. That Koornstra's model is not log-linear is confirmed by $2I(x_K:x_E)$ not equalling the difference $2I(x:x_E) - 2I(x:x_K)$, where x is the observed table, x_K the fitted table under Koornstra's model and x_E the fitted table under the exposure only model.

The fit of the model with exposure alone, a simple independence model, is good. The improvement in fit obtained by going to Koornstra's model involving both exposure and proneness is not significant at the 15 percent level, by reference to the distribution of a chi-squared with 8 degrees of freedom. Now the effect of differences in proneness manifest themselves as deviations from the simple independent model, leading one to expect the estimates

TABLE 5

EM Fits of Koormstra's Model

Parameter Estimates

Code:		m_{21}	m_{30}	m_{40}	m_{50}	m_{60}	m_{70}	w_{23}	w_{37}	w_{55}
Table										
Original	e_i	6.182	7.662	6.196	3.058	1.610	0.332	0.468	1.657	1.209
	p_i	1.167	0.674	0.479	1.628	2.165	8.350	2.690	0	0.574
Smoothed	e_i	5.897	7.943	6.375	3.014	1.805	0.653	0.637	1.572	1.258
	p_i	1.268	0.615	0.453	1.708	1.851	4.151	2.179	0.243	0.745

Goodness of fit	X^2	MDI Statistic
Original Table	25.5589	25.7696
Smoothed Table	27.6345	25.6925

on 28 degrees of freedom

TABLE 6

Goodness of Fit Summary

Statistic	Original Table	Smoothed Table	Degrees of Freedom
$2I(x{:}x_K)$	25.7696	25.6925	28
$2I(x{:}x_E) - 2I(x{:}x_K)$	11.8766	7.9078	8
$2I(x{:}x_E)$	37.6462	33.6004	36
$2I(x_K{:}x_E)$	12.4237	7.7645	8
$X^2(x{:}x_K)$	25.5589	27.6345	28
$X^2(x{:}x_E)$	34.6330	34.2773	36

x: the table
x_K: fitted table under Koornstra's model
x_E: fitted table with the Exposure only
independence model.

of the p_i's in Koornstra's analysis to be relatively unstable. Also the method of fitting he uses seems not to be as efficient as the one used here, simply on the basis of his chi-squared statistic, which is larger for his fit of his model than it is for the exposure only model fit via the MDI approach.

These results are from the analysis of a single table, and cannot be extrapolated without further verification. They do, however, raise questions about the validity of models involving both exposure and proneness parameters, and illustrate the necessity of first rejecting simple models in this type of research.

REFERENCES

Dempster, A.P., N.M. Laird and D.B. Rubin, 1977: "Maximum Likelihood from Incomplete Data via the EM Algorithm", Journal of the Royal Statistical Society (Series B), vol. 39, No. 1, pp. 1-38.

Koornstra, M.J., 1973a: "A Model for Estimation of Collective Exposure and Proneness from Accident Data", Accident Analysis and Prevention, Vol. 5, pp. 157-173.

Koornstra, M.J., 1973b: "Empirical Results on the Exposure-Proneness Model", Accident Analysis and Prevention, Vol. 5, pp. 175-189.

THE MINIMUM DISCRIMINATION INFORMATION ANALYSIS OF CONTINGENCY TABLES IN TRAFFIC SAFETY STUDIES

D.V. Gokhale

Department of Statistics, University of California, Riverside, California 92521, U.S.A.

SUMMARY

This paper deals with analysis of frequency data from traffic safety investigations by using the minimum discrimination information approach. It is shown how the present method is more appropriate and how it often provides deeper insight into the phenomenon under study, as compared to the usual methods such as regression analysis, analysis-of-variance or two-way-chi-square.

1. INTRODUCTION

This paper is concerned with the analysis of frequency or count data that arise in traffic safety investigations. Methods of analysis presented here offer more valid means of interpreting the data. In many instances, they are more appropriate than classical methods and provide a deeper insight into the problem. These statements are illustrated here with the help of an interesting example.

Basically, the methodology makes use of the principle of minimum discrimination information (MDI) for model building and model validation in multinomial experiments. A particular case of the model building aspect is that of "Log-linear models". Application of these models in multidimensional contingency tables is discussed by Oppe (1980), Andersen (1980) and others. The present paper uses a different parameterization and points out some interesting interpretations.

There are essentially three standard methods of analyzing discrete data: (i) two-way chi-square, (ii) simple or multiple regression, and (iii) analysis of variance. The two-way chi-square often provides some insight into the problem but gives only a pairwise analysis. It completely ignores the effect of other variables that are present and does not permit analysis of three or higher factor interactions. Regression and analysis of variance methods require some or all of the following assumptions for a complete analysis of actual measurements:

(i) Random sampling;

(ii) A structure on the population means (such as

$E(Y_{ij}) = \mu + \alpha_i + \beta_j)$;

(iii) Constancy of variance ($\mathrm{Var}(y_{ij}) = \sigma^2$ for all i and j);

(iv) Normality of underlying distributions.

Without assumption (iv), validity of the F-tests and other tests is open to doubt. As a remedy, some authors use "distribution-free" procedures. But for these too, assumptions such as continuity of distribution functions are required. Another usual remedy is the use of transformations. But after analyzing the transformed data it often becomes difficult to interpret the results as they relate to the actual measurements. Further, in that process the statistical properties of estimates may become intangibly warped.

The methodology proposed here does away with all the assumptions above *except* (i). On the other hand, there are hardly any applications-oriented statistical techniques which can provide something significant without that assumption. Since the analysis requires minimal assumptions the conclusions are valid for a wider class of problems. This validity is obtained at the cost of the requirement that the *sample be large*. As the following examples show, there are many studies where paucity of data is not a problem. Also when frequencies are analyzed rather than actual measurements, one immediate advantage is that the underlying variable can be quantitative as well as qualitative.

This paper aims at providing more appropriate substitutes for regression and analysis of variance methods. In the following two sections the necessary notation and background is developed. Mathematical details are kept to a minimum. They can be found in the monographs by Kullback (1959) or Gokhale and Kullback (1978). For other models such as those based on "logits", the reader is referred to Bishop et al. (1975), Fienberg (1977), or Gokhale and Kullback (1978). Also, methods of handling (too many) zero counts are not discussed here, their nature being rather technical. They may be found in the monographs mentioned above. The MDI approach was used to analyze many data sets by several groups of participants of the NATO Advanced Study Institute held at Sogesta, Italy, in June 1979. Illustrative examples in this

paper are based on problems discussed by these participants at
the Institute.   Their permission to do so is gratefully acknowledged.

2.   NOTATION AND PRELIMINARIES

Consider N observations classified independently into a certain number (denoted by Ω) of cells according to one or more *char-
actersitics*.   Each characteristic may have several different *categories* (or *levels*) of classification.   Data represented in such
grouped forms are called frequency distributions.   An important
special case of frequency distribution is a *contingency table*.
The methodology described in this paper is applicable to all forms
of tabular data for which the table-entries are frequencies.   The
table may have one or several dimensions of classification.   For
contingency tables the conventional notation "4 × 2 × 5" signifies
that it is a three-way table, the first characteristic has four
categories, the second has two and the third had five.   (In this
instance $\Omega = 40$.)   More details are supplied by the table of
characteristics illustrated in the example below.

A typical frequency of occurrence in the body of the table is
denoted by the symbol x followed by the cell index in parentheses,
$x(hij)$.   The corresponding observed proportions are denoted by
$\hat{\pi}(hij)$ with $\hat{\pi}(hij) = x(hij)/N$.   The underlying probabilities are
denoted by $p(hij)$.   Thus $p(hij)$ is the probability than an individual observation is classified in the cell (hij).   Note that
the $p(hij)$ are constant from one individual to the other.   In many
instances cell (hij) is associated with certain meaningful numbers
which are denoted by symbols such as $a(hij)$, $v(hij)$, etc..
The *dot notation* is used to denote totals.   Thus

$$x(\cdots) = \sum_h \sum_i \sum_j x(hij) = N,$$

$$\hat{\pi}(h{\cdot}j) = \sum_i \hat{\pi}(hij),$$

$$p(\cdot i \cdot) = \sum_h \sum_j p(hij),$$

and so on.

For a two-way table the cell probabilities are denoted by
$p(hi)$, for a three-way table by $p(hij)$, etc..   It is extremely
convenient to use a unified notation $p(\omega)$, say, where ω generic-
ally denotes an (hi) cell in a two-way table, an (hij) cell in a
three-way table, etc..   This is achieved by arranging the cells
of a contingency table in *lexicographic* order.   For example, in a
$4 \times 2 \times 3$ table the symbol ω takes values $1,2,\ldots,24$ and the cor-

respondence between ω and the cell index is given by

$$\begin{array}{lccccccc} \omega: & 1 & 2 & 3 & 4 & \cdots & 24 \\ \text{Cell Index:} & 111 & 112 & 113 & 121 & & 423. \end{array}$$

As an example consider Table 1 of number of motor-cycle accidents classified according to age of rider and sex. The data are fictitious. The characteristics and their levels are tabulated as follows:

Characteristic	Index	Values			
		1	2	3	4
Age in Years	h	15-20	20-25	25-35	35-65
Sex	i	male	female		

Table 1 also provides midpoints of age intervals in square brackets and exposure in vehicle miles (thousands) travelled for each cell in parentheses. This is a 4 × 2 table with Ω = 8. The symbol

TABLE 1

Distribution of Motorcycle Accidents

Age of Rider	Sex		Total
	Male	Female	
15-20 [17.5]	14(16)	11(6)	25
20-25 [22.5]	15(13)	5(8)	20
25-35 [30.0]	20(32)	11(26)	31
35-65 [50.0]	21(13)	9(5)	30

$x(32)$ denotes the frequency 20 in the cell with $h = 3$, $i = 2$. Also $x(\cdot i) = x(11) + x(21) + x(31) + x(41) = 14 + 15 + 20 + 21 = 70$, $\hat{\pi}(3\cdot) = \hat{\pi}(31) + \hat{\pi}(32) = \frac{20}{106} + \frac{11}{106} = \frac{31}{106}$, and so on. The variable age is quantitative and each level can be associated with its midpoint if necessary. Each cell (hi) can then be associated with two numbers, say $a(h)$ and $v(hi)$, $a(h)$ denoting the midpoint of the corresponding age interval and $v(hi)$ denoting the exposure. For example, for cell (42), the two associated numbers are $a(4) = 50$ and $v(42) = 5$, for cell (41) they are 50 and 13 respectively.

In the process of data analysis the underlying probabilities are usually required to satisfy some *linear constraints*. These constraints are expressed in matrix notation as

$$Cp = \vartheta, \tag{1}$$

where C is a $(r + 1) \times \Omega$ matrix and ϑ is a $(r + 1) \times 1$ vector. Both C and ϑ are assumed to be known. The first row of C consists of all ones and the first element of ϑ is also one. This incorporates in the equations the *natural* constraint

$$\sum_{\omega} p(\omega) = 1.$$

As a simple example, suppose that in a 2×2 table the underlying probabilities $p(ij)$ are required to satisfy the same marginal totals as the observed proportions. These constraints can be formulated as

$$p(1\cdot) = \hat{\pi}(1\cdot) \quad \text{(This implies } p(2\cdot) = \hat{\pi}(2\cdot)),$$
$$p(\cdot 1) = \hat{\pi}(\cdot 1) \quad \text{(This implies } p(\cdot 2) = \hat{\pi}(\cdot 2)), \tag{2}$$

and are expressible as (1) by letting

	(11)	(12)	(21)	(22)
$C =$	1	1	1	1
	1	1	0	0
	1	0	1	0

$p = (p(11),p(12),p(21),p(22))'$,

$\vartheta = C\hat{\pi}$,

$\hat{\pi} = (\hat{\pi}(11),\hat{\pi}(12),\hat{\pi}(21),\hat{\pi}(22))'$.

It is assumed throughout for the sake of simplicity and without loss of generality that the rows of C are linearly independent, i.e., rank$(C) = r + 1$. (Hence in the above 2×2 example the constraints $p(2\cdot) = \hat{\pi}(2\cdot)$ and $p(\cdot 2) = \hat{\pi}(\cdot 2)$ should *not* be included in addition to (2)). As another example, suppose that in a 2×2 table the underlying probabilities are required to satisfy the constraint $p(12) = p(21)$ (hypothesis of symmetry), then the matrix

C and vector ϑ of (1) become

$$
\begin{array}{cccc}
(11) & (12) & (21) & (22) \\
\end{array}
$$

$$
C = \begin{array}{cccc}
1 & 1 & 1 & 1 \\
0 & 1 & -1 & 0
\end{array} \quad ,
$$

$$\vartheta = (10)'.$$

The two examples given above also illustrate two important types of problems that arise in the analysis of multinomial experiments. In the first example, the underlying probability distribution is required to have the same marginal totals as the observed distribution. In other words the underlying model is postulated to use that much (and no more) information from the data as given by the specified (and implied) marginal totals. More generally, when linear combinations of underlying probabilities are required to have the same values as given by the same linear combinations of observed proportions, the problem is called an Internal Constraints Problem (ICP). ICP can be looked upon as a *model-building* or smoothing-cum-fitting process. Note that in ICP, the vector ϑ of (1) is obtained "internally", i.e., by using the observed distribution; in fact $\vartheta = C\hat{\pi}$.

In the second example, it was possible to derive the constraint-equations by using the hypothesis under question, i.e., "externally" to the data. Such problems, which arise in *model-validation* are called External Constraints Problems (ECP).

3. THE MDI APPROACH

Let $\tilde{\Omega}$ denote the set of elements $\{1,2,3,\ldots,\Omega\}$. For $\omega \in \tilde{\Omega}$ if $p(\omega)$ and $\pi(\omega)$ are any two probability distributions defined on $\tilde{\Omega}$, the discrimination information $I(p:\pi)$ in p with respect to π is given by

$$I(p:\pi) = \sum_{\omega} p(\omega)\ln\frac{p(\omega)}{\pi(\omega)} \, .$$

(It is assumed that $p(\omega) = 0$ whenever $\pi(\omega) = 0$ and $0\ln 0 = 0$). $I(p:\pi)$ can be looked upon as a measure of closeness between the distributions p and π. For example, if $\bar{\Omega} = \{1,2,3,4\}$

$$\pi(\omega) = \tfrac{1}{4} \text{ for } \omega \in \bar{\Omega},$$

$$p_a(1) = 0.3, \quad p_a(2) = 0.3, \quad p_a(3) = 0.2, \quad p_a(4) = 0.2,$$

and

$$p_b(1) \times = 0.4, \quad p_b(2) = 0.3, \quad p_b(3) = 0.2, \quad p_b(4) = 0.1,$$

then $I(p_a:\pi) = 0.02014$ and $I(p_b:\pi) = 0.1064$ indicating that the distribution p_a is closer to π than p_b. In fact $I(p:\pi)$ is always non-negative and equals zero if and only if $p(\omega) = \pi(\omega)$ for *all* $\omega \in \bar{\Omega}$. Suppose that π is a known "reference" distribution, not necessarily $\hat{\pi}$, and that p is required to satisfy linear constraints of the form $Cp = \vartheta$ given by (1). Consider the problem of choosing from all probability distributions satisfying (1), the one which is closest to π in the sense that it minimizes $I(p:\pi)$. The solution to this problem is known as the minimum discrimination information theorem (Kullback (1959)) which states that such a p^* which minimizes $I(p:\pi)$ subject to (1) exists uniquely under mild regularity conditions and is given by

$$\ln \frac{p^*(\omega)}{\pi(\omega)} = L + \sum_{j=2}^{r+1} C(j,\omega)\tau_j,$$

where the parameters L and τ_j, $j = 1,\ldots,r$, are determined so that p^* satisfies (1). The corresponding MDI estimate of the cell-frequency is denoted by $x^*(\omega) = Np^*(\omega)$.

Internal Constraints Problems: As mentioned earlier, in ICP the constraints are taken as linear combinations of observed proportions. The objective is to smooth out the data by arriving at a model which explains the data in terms of as few parameters as possible. Thus, the reference distribution π may be any distribution which satisfies (1) and does not involve any more parameters than the ones in the respresentation (3). It turns out that the actual choice of such π is immaterial; the estimates of cell frequencies, test-statistics and related analysis remain the same (see Gokhale and Kullback (1978)). The uniform distribution over $\bar{\Omega}$ is the most obvious choice for π in ICP since it only has a scaling parameter. On the other hand the observed distribution $\hat{\pi}$ cannot be used as π in ICP.

The MDI statistic for ICP is $2I(x:x^*)$ which allows one to test the goodness-of-fit. It is distributed as a chi-square with $\Omega - r - 1$ degrees of freedom for large sample sizes N. Further, if: (i) x_a^* is a vector of MDI estimates corresponding to a set of constraints C_a, (ii) x_b^* is such a vector corresponding to constraints C_b, and (iii) the rows of C_a are explicitly or implicitly (as linear combinations) contained in the rows of C_b, we have

$$2I(x:x_a^*) = 2I(x_b^*:x_a^*) + 2I(x:x_b^*).$$

Each term on the right is also an MDI statistic distributed like a chi-square for large samples. The property (4) is called *analysis of information*.

External Constraints Problems: Here the constraints are provided by the hypothesis under question and the problem is to determine whether the departure of $\hat{\pi}$ (if any) from (1) can be attributed to chance. This is achieved by finding the MDI estimate p^* which satisfies (1) and is closest to $\hat{\pi}$ by minimizing $I(p:\hat{\pi})$. Validity of the external hypothesis is tested by the MDI statistic $2I(x^*:x)$, distributed as chi-square with r degrees of freedom in large samples. (Note the difference in the form of the MDI statistics for ICP and ECP and also in their degrees of freedom).

Analysis of information can also be applied in ECP. If an external hypothesis formulated as $C_2p = \vartheta_2$ implies another external hypothesis $C_1p = \vartheta_2$ where C_2 is $(r_2 + 1) \times \Omega$ and C_1 is $(r_1 + 1) \times \Omega$ with $r_2 > r_1$,

$$2I(x_2^*:x) = 2I(x_2^*:x_1^*) + 2I(x_1^*:x).$$

The L.H.S. has r_2 degrees of freedom, the first and second terms on the R.H.S. have $r_2 - r_1$ and r_1 degrees of freedom respectively.

Estimates of Covariances: The representation (3) can itself be looked upon as the underlying probability distribution with L and the τ_j being the unknown parameters. Their values calculated from the data are estimates and are hence subject to chance errors. Usually, the covariances of the estimates of tau-parameters are of interest. The covariances are estimated as a matrix denoted by $(S_{22.1})^{-1}$ where

$$S_{22.1} = S_{22} - S_{21}S_{11}^{-1}S_{12},$$

$$S = \left[\begin{array}{c|c} S_{11} & S_{12} \\ \hline S_{21} & S_{22} \end{array} \right], \text{ a partitioned matrix where}$$

S_{11} is 1×1, S_{22} is $r \times r$, $S_{12} = S_{21}'$ is $1 \times r$.

$$S = CDC',$$

where C is given in (1) and D is a diagonal matrix with $x^*(\omega)$ in the ω-th diagonal position.

Outliers: In internal constraints problems it happens on some occasions that the model fits the data very well *except* for one or two "outlier" cells. These outliers lead to rejection of the model on an overall basis due to large contributions to the goodness-of-fit MDI-statistic $2I(x:x^*)$. For each cell, it is possible to find a lower bound to the contribution of the cell-frequency to $2I(x:x^*)$. These lower bounds are called OUTLIER values and can be used effectively in model search (see Gokhale and Kullback (1978, p. 64)). If there are only one or two cells with large outlier contributions to $2I(x:x^*)$ they often indicate some errors such as in coding or punching or some other feature of the data.

Use of outliers is illustrated in one of the examples to follow.

Percentage "Variation" Explained: Consider an ICP. From the analysis of information (4) it is clear that the goodness-of-fit MDI statistic $2I(x:x_a^*)$ is decomposed into two components. One, $2I(x:x_b^*)$, is the goodness-of-fit statistic corresponding to model (b) and the other $2I(x_b^*:x_a^*)$ is a measure of the effect of using more information from the data, in terms of constraints added in C_b. The ratio

$$\frac{2I(x:x_a^*) - 2I(x:x_b^*)}{2I(x:x_a^*)} = \frac{2I(x_b^*:x_a^*)}{2I(x:x_a^*)}$$

can be looked upon as a fraction of the original "variation" (disparity between the data and model (a)) explained by using model (b) which is more complex (has more parameters) than model (a). A similar interpretation holds for ECP.

The Case of Several Samples: It is easily possible to extend the foregoing methodology to the case of more than one sample. But the notation becomes clumsy and the discussion a little more technical. The reader is referred to Gokhale and Kullback (1978) for the general treatment. Some indication of the underlying theory is given as an example in Section 4.

Computer Programs: The computer programs necessary for application of the MDI approach to frequency data are written in PL-1 with optimizing compiler. They are working at least at the George Washington University Computer Center and at the Computer Center of the University of California, Riverside. Interested readers should contact the Department of Statistics, George Washington University, Washington, D.C. 20052, where tapes of programs can be made available at a nominal charge.

4. ILLUSTRATIVE EXAMPLES

In this section we give three examples to illustrate the foregoing methodology. The first example reanalyzes a certain data set as ICP using log-linear models. It illustrates the use of OUTLIER values in building appropriate models and also shows how the model can be expressed in terms of parameters representing various associations. The second example builds models that make use of exposure data in ICP. The third example treats the problem of comparing two binomial proportions as a simple case of an ECP with several samples. For the second and third examples actual computer output has not been obtained. The purpose of the examples is to explain how to formulate the problem rather than to give the actual one its analysis.

4.1 Example 1

The data of this example deal with rear-end damage to heavy goods vehicles (Table 4.1.1). The study, described below, was undertaken by the Transport and Road Research Laboratory, Crowthorne, England. The author regrets the unavailability of more detailed references at the time of preparation of the manuscript of this paper.

TABLE 4.1.1

Data on Damage to Heavy Goods Vehicles

(See table of characteristics and levels for notation of cell index)

Cell Index *hijku*	Observed Frequency *x*	Cell Index *hijku*	Observed Frequency *x*	Cell Index *hijku*	Observed Frequency *x*
11111	405	21111	540	31111	68
11112	366	21112	349	31112	50
11121	119	21121	80	31121	8
11122	110	21122	48	31122	7
11211	307	21211	94	31211	277
11212	247	21212	62	31212	129
11221	73	21221	15	31221	38
11222	69	21222	7	31222	32
12111	1317	22111	265	32121	113
12112	1194	22112	227	32122	88
12121	5171	22121	996	32121	183
12122	4525	22122	755	32122	126
12211	1240	22211	60	32211	466
12212	1179	22212	56	32212	406
12221	5578	22221	260	32221	835
12222	5243	22222	232	32222	759

Analysis using two-way chi-square and other techniques developed by Nelder and Wedderburn (1972) was reported by Miss Patricia McBean at the Sogesta NATO-ASI.

Here the data are reanalyzed using the MDI approach. There

are five characteristics giving rise to a $3 \times 2 \times 2 \times 2 \times 2$ contingency table having 48 cells ($\Omega = 48$). The analysis is of an exploratory nature; we want to build a model that is as simple as possible, retains pertinent and meaningful associations amongst the variables under study and gives a good fit to the data. Hence we proceed with the ICP treatment. The analysis is started by first preparing a table of characteristics, their levels and indexes used.

Characteristic	Index	Values		
		1	2	3
Light Condition	h	Day	Dark lit	Dark unlit
HGV Parked	i	Yes	No	
Location	j	Urban	Rural	
Damage	k	Rear-end	Other	
Time Period	u	1969-71	1971-73	

The aim of the study was to examine the effect on rear-end damage, of the law passed in 1971, requiring HGVs to use strip markers visible at night. If the law does have an effect the proportion of cases with rear-end damage would be reduced significantly in the second time period. Other characteristics are concomitant and control for changes from one time period to another.

To see whether there is a significant change in the distribution of cases with rear-end damage over combinations of levels of other variables is different for the two periods. We start with the *base* null hypothesis that the two distributions are the same. Under this hypothesis, the common distribution is estimated by $\hat{\pi}(hijk\cdot)$, h = 1,2 or 3, i = 1 or 2, j = 1 or 2 and k = 1 or 2. The proportion of cases in 1969-71 is estimated by $\hat{\pi}(\cdots\cdot1)$ and that in 1971-73 by $\hat{\pi}(\cdots\cdot2)$. Under the hypothesized model (a) the distributions for the two time periods are the same, the MDI estimate $p_a^*(hijku)$ is given by

$$p^*(hijku) = \hat{\pi}(hijk\cdot)\hat{\pi}(\cdots\cdot u).$$

In terms of the ICP formulation, hypothesis (a) can be interpreted by saying that the MDI estimate uses the four-way observed marginals $x(hijk\cdot)$ and the one-way observed marginals $x(\cdots\cdot u)$. No other information from the data is needed to give a good fit. Whether this hypothesis is tenable is determined by the goodness-of-fit MDI statistic $2I(x:x_a^*)$. Note that we have started with four-way marginals $x(hijk\cdot)$ to retain, from the data, associations of all order among the first four variables of interest. If model

(a) is valid, $2I(x:x_a^*)$ is distributed like a chi-square with 23 degrees of freedom (D.F.). The calculated value of $2I(x:x_a^*)$ is 102.30, too large to accept model (a), showing that there is a strongly significant change in the distributions of cases from 1969-71 to 1971-73. The next question is to locate the variable of interest in which an appreciable change has occurred. The variable of *a priori* interest being rear-end damage, we add pairwise association parameters between damage and time (denoted by τ_{ku}^{DE}) to model (a), call it model (b) and see whether the latter provides a good fit. If it does, one can say that the disparity in the two distributions can be completely explained by changes in rear-end damage cases over the two time periods. The MDI estimates $x_b^*(hijku)$ are obtained by fitting the observed marginals $x(hijk\cdot)$ as in model (a) and the marginals $x(\cdots ju)$. Lower order marginals $x(\cdots j\cdot)$ and $x(\cdots\cdot u)$ are fitted automatically. The goodness-of-fit MDI statistic $2I(x:x_b^*)$ is calculated to be 97.26 with 22 D.F. which is highly significant. The percentage "variation" explained is only 4.92%. This shows that a better model may be obtained by including associations of other variables with time. Before looking for one, we examine the question whether there is a significant change in the variable "damage" over the two time periods. This question is answered by analysis of information. The difference MDI statistic $2I(x:x_a^*) - 2I(x:x_b^*) = 2I(x_b^*:x_a^*)$ tests whether the association parameters τ_{ku}^{DE} can be taken as zero. Its value is 102.30 - 97.26 = 5.04 with 1 D.F. which is significant. The conclusion so far is that a satisfactory model should include association parameters of other variables with time and should, at the same time, explain the significant disparity between models (a) and (b).

Since light condition is another variable of interest a parallel analysis for that variable is carried out. Model (c) adds light condition x time associations to model (a). The goodness-of-fit MDI statistic $2I(x:x_c^*) = 62.41$ with 21 D.F. shows that inclusion of light condition x time associations alone does not provide a good fit to the data. On the other hand the difference MDI statistic $2I(x_c^*:x_a^*) = 39.89$ with 2 D.F. points out the strong significance of the effects of these associations.

Analyses of models (b) and (c) show that *both* pairwise associations, damage x time and light condition x time, are significant *when tested separately*. However, since light condition and damage may themselves have some association it is interesting to see whether inclusion of, say, damage x time associations is necessary *given that* light condition x time associations are already included in the model. This aspect is similar to the one of "partial correlations" in continuous multivariate analysis. We would like to "adjust" for the effect of change in light condition in studying the effect of change in rear-end damage over time. We thus fit model (d) which includes both pairwise associations, damage x

time and light condition x time and compare it with model (c)
which has light condition x time associations only. The difference $2I(x_d:x_c^*) - 2I(x:x_c^*) = 2I(x:x_d^*) = 1.470$ with one D.F. and
measures the effect of inclusion of damage x time associations
given that light condition x time association parameters are already included in the model. The value 1.470 is not significant
even at 25% level showing that model (c), which accounts for
changes in light conditions, also "explains" changes in rear-end
damage cases indirectly. On the other hand, $2I(x_d^*:x_b^*) = 2I(x:x_b^*)$
$- 2I(x:x_d^*) = 97.26 - 60.94 = 36.32$ with $22 - 20 = 2$ D.F. is very
highly significant showing that model (b) does not account for
changes in light conditions.

At this stage it should be remembered that model (d) has not
provided a good fit to the data, $2I(x:x_d^*)$ being 60.94 with 20 D.F..
It has only assessed the importance of τ_{ku}^{DE} given that parameters
τ_{hu}^{AE} are in the model (model (c)). To look for a better model we
go back to model (c) and examine the OUTLIER values. There are
two outlier values corresponding to cells (31211) and (31212).
Hence cells corresponding to dark unlit light condition ($h = 3$)
for parked HGVs ($i = 1$) in rural locations ($j = 2$) show a markedly
different association pattern in rear-end versus other damage
over time. We therefore try a model, say model (e) , in which
the observed values $x(31211)$ and $x(31212)$ are excluded from the
smoothing process. In other words, we use a model that fits all
the marginals included in model (c) with the additional constraint
$x_e^*(31211) = x(31211)$ implying automatically that $x_e^*(31212) =$
$x(31212)$. The MDI estimates $x_e^*(hijku)$ and the tau-parameters of
model (e) are different from those of model (c). Under model (e)
$2I(x:x_e^*) = 34.60$ with 20 D.F. which is not significant at any
reasonable level, shows a good fit. There are no outliers. Unfortunately the size of the data set was too large for the available computer to provide estimates $x_e^*(hijku)$ in the KULLITR output.

If estimates of tau-parameters and estimates of their covariance matrix are required for further investigation, they can be
obtained from the output of KULLITR program. Here we list the
28 parameters included in model (e). The 20 D.F. for this model
correspond to the remaining 20 parameters assumed to be zero in
(or absent from) model (3). The log-linear model is obtained by
equating $\ln(p_e^*(hijku)/\pi(hijku))$, π being the uniform distribution
over 48 cells, to the sum of the following τ-parameters: marginals
fitted: $x(hijk\cdot)$, $x(h\cdot\cdot k)$, $x(31211)$.

Scaling Constant (1) L;

One-Way Marginals (6)
$h \quad \tau_1^A \tau_2^A, \qquad i \quad \tau_1^B, \qquad j \quad \tau_1^C, \qquad k \quad \tau_1^D, \qquad u \quad \tau_1^E;$

Two-Way Marginals (11)

hi $\tau^{AB}_{11}, \tau^{AB}_{21}$

hj $\tau^{AC}_{11}, \tau^{AC}_{21}$

hk $\tau^{AD}_{11}, \tau^{AD}_{21}$

hu $\tau^{AE}_{11}, \tau^{AE}_{21}$

ij τ^{BC}_{ij}

ik τ^{BD}_{ik}

jk τ^{CD}_{11}

Three-Way Marginals (7)

hij $\tau^{ABC}_{111}, \tau^{ABC}_{211}$

hik $\tau^{ABD}_{111}, \tau^{ABD}_{211}$

hjk $\tau^{ACD}_{111}, \tau^{ACD}_{211}$

ijk τ^{BCD}_{111}

Four-Way Marginals (2)

$hijk$ $\tau^{ABCD}_{1111}, \tau^{ABCD}_{2111}$

Cell (31211) τ^{ABCDE}_{31211}

It is easy to list these tau-parameters by noting the marginals fitted, $x(hijk\cdot)$, $x(h\cdot\cdot\cdot u)$ and $x(31211)$. Hence association (tau) parameters of all order between $(hijk)$, univariate and pairwise parameters between (hu) and one parameter corresponding to cell (32121) are to be included in the model.

The overall conclusions of our analysis of this data set are that (i) the data do not seem to show a significant effect of passing the law requiring strip markings, (ii) changes in light conditions over the two time periods account for changes in cases of rear-end damage and (iii) incidence of rear-end damage for HGVs parked on dark unlit roads in rural areas calls for further investigation.

4.2 Example 2

In traffic studies the number of vehicle-miles travelled is often an important ancilliary variable. The present example illustrates model-building (ICP) by making use of available exposure data. Table 4.2.1 gives fictitious data on number of accidents to bicycle riders classified according to degree of urbanization $-i-$ and age of the cyclist $-j$. In the same table, the number, $V(ij)$, of driven kilometers in milliards is given in parentheses. The objective is to build a model that incorporates the information provided by $V(ij)$.

TABLE 4.2.1

Number of Accidents to Bicycle Riders

(Driven Kilometers in milliards are in parentheses)

Urbanization	Age		
	16-20	21-60	61 and over
Capital	127 (0.08)	257 (0.23)	118 (0.13)
Suburbs	38 (0.03)	39 (0.20)	38 (0.06)
Towns	7 (0.01)	423 (0.01)	327 (0.05)
Rural	15 (0.05)	113 (0.05)	11 (0.03)

Since the $V(ij)$ themselves depend on urbanization and age of rider, the simplest model is obtained by assuming that (a) given $V(ij)$ the underlying probabilities $p(ij)$ do not depend on i or j (see Figure 4.2.1) and, (b) further, the model is linear in $V(ij)$. We then get

$$\ln[p(ij)/\pi(ij)] = L + \alpha V(ij),$$

where $\pi(ij)$ is the uniform distribution over 12 cells and L and α are unknown parameters.

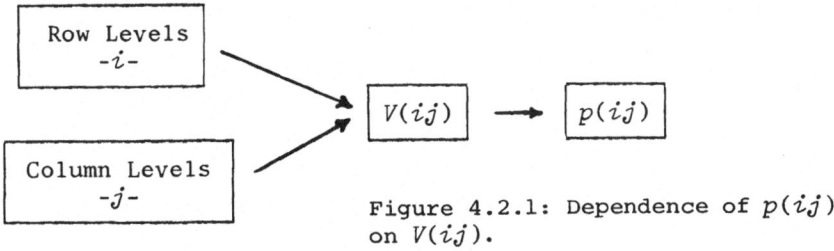

Figure 4.2.1: Dependence of $p(ij)$ on $V(ij)$.

This being an ICP, the MDI estimates $p*(ij)$ are obtained by minimizing $I(p:\pi)$ subject to $Cp = C\hat{\pi}$ where matrix C is

$$
\begin{bmatrix}
1 & 1 & 1 & 1 & \cdots & 1 \\
v(11) & v(12) & v(13) & v(21) & \cdots & v(43)
\end{bmatrix} .
$$

Test of the above model is provided by the MDI statistic $2I(x:x*)$ with $12 - 2 = 10$ D.F..

If the model does not give a good fit there are several ways to build a more complex model. One is to keep the assumption (a) above and include quadratic and higher order terms in $V(ij)$. Another one, which we prefer, is to retain assumption (b) and assume that the observed row and column marginals contain some information about the $p(ij)$ as a direct input (see Figure 4.2.2).

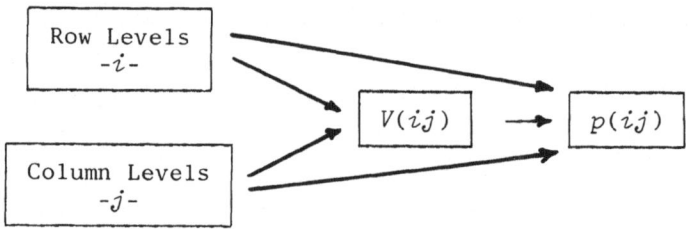

Figure 4.2.2: Dependence of $p(ij)$ on $V(ij)$ and (ij).

Then a simple model is

$$
\ln \frac{p(ij)}{\pi(ij)} = L + V(ij) + \tau_i^A + \tau_j^B,
$$

where A and B denote the row and column variables respectively. The corresponding C matrix for the 4×3 example is given below:

Cell Index

11	12	13	21	22	23	31	32	33	41	42	43
1	1	1	1	1	1	1	1	1	1	1	1
$v(11)$	$v(12)$ •	•	•	•	•	•	•	•	•	•	$v(43)$
1	1	1	0	0	0	0	0	0	0	0	0
0	0	0	1	1	1	0	0	0	0	0	0
0	0	0	0	0	0	1	1	1	0	0	0
1	0	0	1	0	0	1	0	0	1	0	0
0	1	0	0	1	0	0	1	0	0	1	0

The goodness of fit MDI statistic $2I(x:x^*)$ now has $12 - 7 = 5$ D.F..
Difference between MDI statistics of the two models measures the
effect of adding the 5 rows to the first C-matrix, corresponding
to dependence of the $p(ij)$ on the observed marginals $x(i\cdot)$ and
$x(\cdot j)$. Actual numerical analysis (not presented) can be carried
out by using KULLITR program.

A more sophisticated analysis of the present data set can be
done by using the fact that the variable age is quantitative. For
this purpose the reader is referred to Fienberg (1977) and McCullagh (1978).

4.3 Example 3

This third and final example does not contain any data as it
treats the well known problem of comparing two binomial proportions
using the MDI approach. It compares the MDI solution with a few
of its competitors such as maximum likelihood and introduces the
reader to the k-sample ECP (with $k = 2$).

Consider two independent binomial experiments with N_1 and N_2
observations respectively. Let $p(11)$ and $p(12)$ be the underlying
probabilities of "success" and "failure" respectively for the
first experiment and $p(21)$ and $p(22)$ be these probabilities for
the second experiment. The hypothesis under question is H: $p(11)$
$= p(21)$, which implies $p(12) = p(22)$. For example, suppose that
50 male and 55 female bicycle riders from a certain community are
given a true-false test on bike-safety rules. The hypothesis H:
$p(11) = p(21)$ postulates that there is no difference due to sex
in the performance of the two groups. Here $N_1 = 50$, $N_2 = 55$,
$p(11)$ is the probability that a male rider answers the question
correctly, $p(12)$ is the probability that he answers it incorrectly.
For females these probabilities are denoted by $p(21)$ and $p(22)$ respectively.

Let p be the vector obtained by catenating the probability vectors of each experiment in a sequence. Thus

$$p = (p(11),p(12),p(21),p(22))'.$$

There are two natural constraints in the fact that for each experiment the probabilities add up to unity. The hypothesis $p(11)$
$= p(21)$ is a linear constraint on the elements of p. These constraints are written in the form $Bp = \vartheta$, with the natural constraints written first. The matrix is denoted by B in a k-sample
problem $(k > 1)$ as a convention; the symbol C is reserved for a
one sample problem. Thus

$$B = \begin{matrix} 1 & 1 & 0 & 0 \\ 0 & 0 & 1 & 1 \\ 1 & 0 & -1 & 0 \end{matrix} , \qquad \vartheta = (1,1,0)'.$$

Now, if $(\pi(11),\pi(12))$ and $(\pi(21),\pi(22))$ are two arbitrary but known probability vectors then the information functions for the two samples can be written as

$$I_j(p_j:\pi_j) = p(j1)\ln \frac{p(j1)}{\pi(j1)} + p(j2)\ln \frac{p(j2)}{\pi(j2)} , \qquad j = 1,2.$$

Since the hypothesis is formulated without using any observed values, the example presents an ECP. Then $\pi_j = \hat{\pi}_j$, $j = 1,2$ and the vectors $\hat{\pi}_j$ of observed proportions are assumed to be positive. The MDI principle finds a vector p^* which satisfies the hypothesis (so $Bp^* = \vartheta$) and minimizes a *weighted* combination of the k information functions. The weighted combination is

$$\sum_{j=1}^{k} w_j I_j(p_j:\pi_j), \qquad\qquad (5)$$

where w_j are known weights, $w_j > 0$, $\sum_{j=1}^{k} w_j = 1$. Choice of w_j is arbitrary. For example, w_j may be chosen to indicate the "reliability" (possibly subjective) of experiment j. When such a choice of w_j is not available, it can be taken proportional to the sample size; $w_j = N_j / \sum_{j=1}^{k} N_j$.

Now, let P be a vector obtained by multiplying in p_1 each element of the probability vector p_j by w_j. Let vector Π be defined similarly. Thus, in the present example,

$$P = (w_1 p(11), w_1 p(12), w_2 p(21), w_2 p(22))',$$

and

$$\Pi = (w_1 \pi(11), w_1 \pi(12), w_2 \pi(21), w_2 \pi(22))'.$$

The weighted information function (5) is, in fact, $I(P;\Pi)$ where P and Π are probability vectors. Observe that

$$\sum_{j=1}^{2} w_j I_j(p_j:\pi_j) = w_1 I_1(p_1:\pi_1) + w_2 I_2(p_2:\pi_2)$$

$$= w_1 p(11)\ln \frac{p(11)}{\pi(11)} + w_1 p(12)\ln \frac{p(12)}{\pi(12)}$$

$$+ w_2 p(21)\ln \frac{p(21)}{\pi(21)} + w_2 p(22)\ln \frac{p(22)}{\pi(22)} =$$

(Contd)

(Contd) $= w_1p(11)\ln \dfrac{w_1p(11)}{w_1\pi(11)} \quad w_1p(12)\ln \dfrac{w_1p(12)}{w_1\pi(12)}$

$$+ w_2p(21)\ln \dfrac{w_2p(21)}{w_2\pi(21)} + w_2p(22)\ln \dfrac{w_2p(22)}{w_2\pi(22)}$$

$$= I(p:\Pi).$$

The constraints $Bp = \vartheta$ are expressible as constraints on P by defining a block diagonal matrix W of weights w and writing $Bp = \vartheta$ as $BW^1Wp = BW^{-1}P = \vartheta$. In this example,

$$W = \begin{matrix} w_1 & 0 & 0 & 0 \\ 0 & w_1 & 0 & 0 \\ 0 & 0 & w_2 & 0 \\ 0 & 0 & 0 & w_2 \end{matrix}$$

Letting $C = BW^{-1}$, above constraints become $CP = \vartheta$ and the original k-sample problem becomes a one sample problem of finding the MDI estimate P^* of a probability vector which minimizes $I(P;\Pi)$ subject to $CP = \vartheta$. The MDI estimates p_j^* for each sample can then be found by dividing the appropriate element of p^* by w_j.
   In the present example, the MDI estimates p_2^* and p_2^* can be obtained after some straightforward derivation. They come out to be

$$p^*(11) = p^*(21)$$

$$= \frac{\hat{\pi}(11)^{W1}\hat{\pi}(21)^{W2}}{\hat{\pi}(11)^{W1}\hat{\pi}(21)^{W2} + \hat{\pi}(12)^{W1}\hat{\pi}(22)^{W2}},$$

$$p^*(12) = p^*(22) = 1 - p^*(11).$$

The MDI statistic for testing H: $p(11) = p(21)$ is

$$2NI(p^*:\pi) = - 2N\ln\{\hat{\pi}(11)^{W1}\hat{\pi}(21)^{W2} + \hat{\pi}(12)^{W1}\hat{\pi}(22)^{W2}\},$$

when $w_1 = N_1/(N_1 + N_2)$ and $w_2 = N_2/(N_1 + N_2)$. Under hypothesis H: $p(11) = p(21)$, $2NI(p^*:\Pi)$ is distributed like a chi-square with 1 D.F. for large samples.
   We conclude this example by mentioning a few competitors of the above test. An exact (available for all sample sizes, large as well as small) "uniformly most powerful unbiassed" test is given in Lehmann (1959). It requires use of tables of the hypergeometric distribution. For large samples there are three more competitors. One is the "likelihood ratio" test which can be

obtained equivalently by calculating $2NI(\hat{\pi}:\hat{P})$, where $\hat{P}$ is the maximum likelihood estimate of P given by

$$\hat{p}(11) = \hat{p}(21) = \frac{x(11) + x(21)}{N_1 + N_2} , \qquad (6)$$

$$\hat{p}(12) = \hat{p}(22) = 1 - \hat{p}(11). \qquad (7)$$

For large samples $2NI(\hat{\pi}:\hat{p})$ has a chi-square distribution with one D.F. under H. The other two test statistics are

$$Z = \frac{\hat{\pi}(11) - \hat{\pi}(21)}{\sqrt{\dfrac{\hat{\pi}(11)(1 - \hat{\pi}(11))}{N_1} + \dfrac{\hat{\pi}(21)(1 - \hat{\pi}(21))}{N_2}}} ,$$

and

$$T = \frac{\hat{\pi}(11) - \hat{\pi}(21)}{\sqrt{\dfrac{\hat{p}(11)(1 - \hat{p}(11))}{N_1 + N_2}}} ,$$

with $p(11)$ given by (6). Both Z as well as T are distributed asymptotically normally with mean zero and unit standard deviation under H. Properties of these test statistics is still a topic of discussion in current statistical literature (Robbins, 1977).

REFERENCES

Andersen, E.B. (1980): Contingency Tables, *Proceedings of the NATO-ASI on Contingency Table Analysis for Road Safety Studies held at Sogesta, Italy, June 1979*. Alphen aan den Rijn, The Netherlands: Sijthoff and Noordhoff, pp. 3-34.

Bishop, Y.M.M., Fienberg, S.E. and Holland, P.W. (1975): *Discrete Multivariate Analysis*. Cambridge, Mass., U.S.A.: The MIT Press.

Fienberg, S.E. (1977): *The Analysis of Cross-Classified Data*. Cambridge, Mass., U.S.A.: The MIT Press.

Gokhale, D.V. and Kullback, S. (1978): *The Information in Contingency Tables*. New York, U.S.A.: Marcel Dekker, Inc..

Kullback, S. (1959): *Information Theory and Statistics*. New York, U.S.A.: John Wiley and Sons.

Lehmann, E.L. (1959). *Testing Statistical Hypotheses*. New York, U.S.A.: John Wiley and Sons.

McCullagh, P. (1978). A class of parametric models for the analysis of square contingency tables with ordered categories. *Biometrika*, Vol. 65, No. 2, pp. 413-418.

Nelder, J.A. and Wedderburn, R.W.M. (1972): Generalized linear
    models. *J. Roy. Statist. Soc. Ser. A*, Vol. 135, pp. 370-384.
Oppe, S. (1980): Methods for the analysis of contingency tables
    in road safety research. *Proceedings of the NATO-ASI on Con-
    tingency Table Analysis for Road Safety Studies held at So-
    gesta, Italy, June 1979.* Alphen ann den Rijn, The Nether-
    lands: Sijthoff and Noordhoff, pp. 35-46.
Robbins, H. (1977): A fundamental question of practical statistics.
    The American Statistician, Vol. 31, No. 2, p. 97.

ITERATIVE FITTING PROCEDURES FOR MINIMUM DISCRIMINATION INFORMATION ESTIMATION IN CONTINGENCY TABLE ANALYSIS

JOHN C. KEEGEL

Department of Mathematics, University of the District of Columbia, Washington, D.C., U.S.A.

Iterative fitting procedures are algorithms that generate a series of points, each calculated from the previous one that converge to a solution. One of the most important such algorithms used in contingency table analysis is iterative proportional fitting. The algorithm was first used by Deming and Stephan and appeared in the article, On Least Squares Adjustment of a Sampled Frequency Table When the Expected Marginal Totals are Known, Ann. Math. Stat. 34, 911-34 (1940). The procedure was used to adjust sampled tables so that that had certain predetermined marginals. The statistical properties of the estimates they obtained were not known at the time. It was not until 1968 that Ireland and Kullback (2) explicitly proved that these estimates were BAN.

Let us formulate the problem for a very simple case. Let $p(i,j)$ be the true underlying probability of an observation coming from cell (i,j). Let $p*(i,j)$ be the estimate of $p(i,j)$ that satisfies the relations $p*(i,.)=p(i,.)$ and $p*(.,j)=p(.,j)$ where the dot indicates summation over the indicated subscript. Not only does $p*(i.j)$ satisfy these relations but is obtained by minimizing

$$(1) \quad \sum_i \sum_j p(i,j)\ln(p(i,j)/\pi(i,j)) \equiv I(p:\pi)$$

where π is some known distribution, frequently the uniform. I is Kullback's information number, and the p* that minimizes I is called the minimum discrimination information estimate of p. Kullback's minimum discrimination information theorem states that if p* minimizes $I(p:\pi)$ and also satisfies $p*(i,.)=p(i,.)$ and $p*(.,j)=p(.,j)$ then there exist numbers $a(i)$ and $b(j)$ such that

$$(2) \quad p*(i,j)=a(i)b(j)\pi(i,j).$$

Using (2) and the requirement that p*(i,.) we get

(3) $a(i)=p(i,.)/\{\sum_{j} b(j)\pi(i,j)\}$

    If we only knew the values of the b(j) then we would know the values of the a(i). It is unfortunate, but we don't. Let us guess and assign each b(j) the value 1. Denote this guess by b(1,j), and denote the corresponding a(i) by a(1,i). Thus,

(4) $a(1,i)=p(i,.)/\pi(i,.)$.

Substituting these two values a(1,i) and b(1,j) in (2), we get a first estimate of p*(i,j) which we denote by p(1,i,j). Therefore, we have

(5) $p(1,i,j)=\{p(i,.)/\pi(i,.)\}\pi(i,j)=a(1,i)b(1,j)\pi(i,j)$.

    Clearly, p(1,i,.)=p(i,.) i.e. the estimates satisfy the first constraint; however, there is no guarantee that the second constraint will be satified. Most likely it will not be satisfied. Perhaps by adjustment of the b's we can force the next estimate of p*(i,j) to satisfy the second constraint. Let the second estimate be denoted by p(2,i,j)=a(1,i)b(2,j)π(i,j). We wish that p(2,.,j)=p(.,j). Thus, we have

(6) $p(2,.,j)=b(2,j)\sum_{i} a(1,i)\pi(i,j)$,

or

(7) $b(2,j)=p(.,j)/\{\sum_{i}a(1,i)\pi(i,j)\}=p(.,j)/p(1,.,j)$,

or

(8) $p(2,i,j)=\{p(.,j)/p(1,.,j)\}p(1,i,j)$

as a(1,i)π(i,j)=p(1,i,j).
    Now p(2,.,j)=p(.,j). Of course, there is again no guarantee that the first constraint is satisfied. Indeed, it probably is not. Using the same tactics as before, we define a new a(i), say a(2,i) and a new estimate of p*(i,j), say p(3,i,j)=a(2,i)b(2,j)π(i,j), that satisfies p(3,i,.)=p(i,.). Now this implies that p(i,.)= a(2,i)∑ b(2,j)π(i,j). The following relation also holds.

$$\sum_{j} b(2,j)\pi(i,j)=\sum_{j} b(2,j)\{p(1,i,j)/a(1,i)\}$$

$$= \sum_{j} p(2,i,j)/a(1,i)=\{1/a(1,i)\}p(2,i,.).$$

Thus we see that

(9) $a(2,i)=\{p(i,.)/p(2,i,.)\}a(1,i)$.

Therefore, we have that

(10) $p(3,i,j)=\{p(i,.)/p(2,i,.)\}p(2,i,j)$.

The pattern emerges as follows.

(11) $p(2L+1,i,j)=\{p(i,.)/p(2L,i,.)\}p(2L,i,j)$ for L=0,1,...

$p(2L,i,j)=\{p(.,j)/p(2L-1,.,j)\}p(2L-1,i,j)$ for L=1,2,...

At first glance, this mathematical juggling act seems to accomplish one thing, the fitting of an i-marginal at the expence of a j-marginal and vice versa. Fortunately, Kullback and Ireland have shown that $I(p(L):\pi)\leq I(p(L-1):\pi)$ for every L. This coupled with the fact that I is a non-negative convex function imply that the sequence p(L) converges to a unique limit which of necessity must be the distribution that minimizes $I(p:\pi)$; viz., p*.
It also happens that a(L,i) converges to a(i), and that b(L,j) converges to b(j). This gives us a method of calculating the minimizing distribution that is amenable to a calculator.

While we have discussed a very simple case; there is no problem at all going to more dimensions and more fixed marginals. Consider p(i,j,k,l) as the probability of an observation coming from cell (i,j,k,l). Suppose we seek the distribution p*(i,j,k,l) that minimizes $I(p:\pi)$ subject to the restrictions p*(i,j,.,.)= p(i,j,.,.), p*(i,.,k,.)=p(i,.,k,.), and p*(.,.,k,l)=p(.,.,k,l). The iteration would then be as follows.

(12) $p(3M+1,i,j,k,l)=\{p(i,j,.,.)/p(3M,i,j,.,.)\}p(3M,i,j,k,l)$

$p(3M+2,i,j,k,l)=\{p(i,.,k,.)/p(3M+1,i,.,k,.)\}p(3M+1,i,j,k,l)$

$p(3M+3,i,j,k,l)=\{p(.,.,k,l)/p(3M+2,.,.,k,l)\}p(3M+2,i,j,k,l)$

for M=0,1,2,3,... . The p(L,i,j,k,l) converge to the minimizing p* which by Kullback's minimum discrimination information theorem has the form $p*(i,j,k,l)=a(i,j)b(i,k)c(k,l)\pi(i,j,k,l)$.
Of course, we usually have a contingency table, say n(i,j,k), which we regard as being np(i,j,k) where p(i,j,k)=n(i,j,k)/n and $n=\Sigma\Sigma\Sigma$ n(i,j,k).
It is worth mentioning that the π distribution of the algorithm is usually taken to be the uniform distribution. Let us assume for the remainder of the discussion that this is the case. It then happens that

(13) $I(p:\pi)=I(p:p*)+I(p*:\pi)$

Furthermore, it happens that $2nI(p:\pi)$ is asyptotically distributed as a χ^2 with degrees of freedom equal to the number of cells less one. Also, $2nI(p:p*)$ is distributed as a χ^2 with degrees of freedom equal to the number of cells less the number of parameters needed to represent p* in the minimum discrimination information theorem. Lastly, $2nI(p*:\pi)$ is a χ^2 with degrees of freedom equal to the number of parameters necessary to represent p* less one.

Example 1   Consider the following set of data from DISCRETE MULTI-VARIATE ANALYSIS, Bishop, Fienberg, and Holland, MIT Press 1975, p. 148. The data can be considered a 2 x 2 x 2 contingency table indexed by i,j,k.

$i=\begin{pmatrix}1 \text{ nodular lymphoma}\\2 \text{ diffuse lymphoma}\end{pmatrix}$ $j=\begin{pmatrix}1 \text{ male}\\2 \text{ female}\end{pmatrix}$ $k=\begin{pmatrix}1 \text{ unresponsive to treatment}\\2 \text{ responsive to treatment}\end{pmatrix}$

ijk	observation	ij marginals		ik marginals	
111	1	11.	5	1.1	3
112	4	12.	8	1.2	10
121	2	21.	13	2.1	15
122	6	22.	4	2.2	2
211	12				
212	1				
221	3				
222	1				

30 total

Estimates of x*(i,j,k)

ijk	x(1,i,j,k)	x(2,i,j,k)	x(3,i,j,k)
111	2.5	1.15	1.15
112	2.5	3.85	3.85
121	4.0	1.85	1.85
122	4.0	6.15	6.15
211	6.5	11.47	11.47
212	6.5	1.53	1.53
221	2.0	3.53	3.53
222	2.0	0.47	0.47

It should be noted that the iteration converges in just 3 steps. This is due to the fact that if x*(i,j,k) is to have the ij and ik marginals of x(i,j,k) then it has the explicit representation:

(14) $x*(i,j,k)=\{x(i,j,.)x(i,.,k)\}/x(i,.,.)$

which is also the maximum liklihood estimate of x(i,j,k) under the hypothesis of the independence of sex and response given lymphoma type. Usually the minimum discrimination information estimates do not have a simple explicit representation and hence convergence is not gotten after a finite number of steps.

Let us place the previous example in a different mathematical framework for a moment. Our problem may then be phrased in the language of matrix theory as follows. Let $\underline{x}'=(x(1,1,1),x(1,1,2),$ $x(1,2,1),x(1,2,2),x(2,1,1),x(2,1,2),x(2,2,1),x(2,2,2))$ i.e. $\underline{x}$ is the vector whose components are the cell entries in lexicographic order. We shall refer to $x(1,1,1)$ as $x(1)$, $x(1,1,2)$ as $x(2)$, etc.

$$\text{Let the matrix } \hat{C}=\begin{bmatrix} 1 & 1 & 0 & 0 & 0 & 0 & 0 & 0 \\ 0 & 0 & 1 & 1 & 0 & 0 & 0 & 0 \\ 0 & 0 & 0 & 0 & 1 & 1 & 0 & 0 \\ 0 & 0 & 0 & 0 & 0 & 0 & 1 & 1 \\ 1 & 0 & 1 & 0 & 0 & 0 & 0 & 0 \\ 0 & 0 & 0 & 0 & 1 & 0 & 1 & 0 \end{bmatrix} \text{ so that } \hat{C}\underline{x}=\begin{bmatrix} x(1,1,.) \\ x(1,2,.) \\ x(2,1,.) \\ x(2,2,.) \\ x(1,.,1) \\ x(2,.,1) \end{bmatrix}=N\begin{bmatrix} \theta*(1) \\ \theta*(2) \\ \theta*(3) \\ \theta*(4) \\ \theta*(5) \\ \theta*(6) \end{bmatrix}.$$

At first glance, it seems that $\hat{C}\underline{x}$ almost represents the problem that we have just dealt with except that the marginals $x(1,.,2)$ and $x(2,.,2)$ are neglected. If one examines the relationship $x(1,1,.)+x(1,2,.)=x(1,.,1)+x(1,.,2)$, one can see that if $x(1,1,.)$, $x(1,2,.)$, and $x(1,.,1)$ are known then $x(1,.,2)$ can easily be calculated and thus is redundant. The same is true for $x(2,.,2)$ using different relations. This lack of redundancy in the $\hat{C}$ matrix is referred to as linear independence or we say $\hat{C}$ is of full rank. Using the above terminology we can restate the problem as follows.

Let us find the vector $\underline{x}*$ so that $\hat{C}\underline{x}*=N\theta*$ and simultaneously minimizes $I(\underline{x}:\pi)$. The matrix $\hat{C}$ and the type of parameterization used e.g. $a(i,j)$'s and $b(i,k)$'s used are closely related. It is clear that $\hat{C}$ is not the only matrix that can specify all constraints $x(i,j,.)$ and $x(i,.,k)$. Thus for each matrix there corresponds a different, but equivalent parameterization. Of course, that leads us to ask which parameterization is best. It seems that at least for the construction of matrices, it would be a great help to follow a simple set of rulesto obtain acceptable (linearly independent) matrices without having to figure out which relation follows from what other relations each time. Consider the following matrix C.

$$C=\begin{bmatrix} 1 & 1 & 1 & 1 & 1 & 1 & 1 & 1 \\ 1 & 1 & 1 & 1 & 0 & 0 & 0 & 0 \\ 1 & 1 & 0 & 0 & 1 & 1 & 0 & 0 \\ 1 & 0 & 1 & 0 & 1 & 0 & 1 & 0 \\ 1 & 1 & 0 & 0 & 0 & 0 & 0 & 0 \\ 1 & 0 & 1 & 0 & 0 & 0 & 0 & 0 \end{bmatrix} \text{ Note that } C\underline{x}=\begin{bmatrix} x(.,.,.) \\ x(1,.,.) \\ x(.,1,.) \\ x(.,.,1) \\ x(1,1,.) \\ x(1,.,1) \end{bmatrix}. \text{ One should verify}$$

that these constraints imply and are implied by the previous set specified by $\hat{C}$. This matrix is easier to construct. We first insure that the total is fixed. This corresponds to the first row of the

matrix. We then fix each of the one-way marginals in turn, then
each of the two-way marginals, etc. The one rule to remember is
that if a marginal has any subscript at the highest level of that
factor it is redundant and hence unnecessary. Thus, we see that
no fixed marginals with a subscript of 2 appear_ in the above.

It is of some interest to examine the form that Kullback's
minimum discrimination information estimate can take if we formulate
the problem as above. The theorem tells us that the minimizing $\underline{x}$,
say $\underline{x}^*$, has the following representation.

$$(15) \qquad x^*(i)=\exp(L+\tau(1)c(1,i)+\tau(2)c(2,i)+\ldots+\tau(6)c(6,i))\pi(i)$$

where L is a normalizing factor and the τ's play a role similar to
regression coefficients. If the above representation is given in
product form and each of the factors is identified as a parameter
then our representation will be exactly in the same form as previously given.

GENERALIZED INDEPENDENCE.

The concept of generalized independence appears frequently in
the literature and is simply another way of viewing the case of
fixed marginals. For the sake of simplicity, suppose we have a
probability distribution $p(i,j,k)$ where $i=1,\ldots,I$, $j=1,\ldots,J$, and
$k=1,\ldots,K$. Then by generalized independence, we mean that $p(i,j,k)$
can be expressed as a product of positive functions of proper subsets of the indices. For example $p(i,j,k)=a(i)b(j)c(k)$, or $p(i,j,k)$
$=a(i,j)b(j,k)$, or $p(i,j,k)=a(i,j)b(j,k)c(i,k)$, etc. The object then
is to find out what these 'component' functions look like and to
determine their properties. In each case, let us estimate p by p*
where p* is obtained by minimizing $I(p:\pi)$ where π is taken to be
some know distribution and for a large class of problems is the
uniform distribution; i.e., $\pi(i,j,k)=1/IJK$ and p* is constrained
to satisfy marginal restrictions indicated by the 'component'
function subscripts.

If we hypothesize that $p(i,j,k)=a(i)b(j)c(k)$ then we require
that $p^*(i,.,.)=p(i,.,.),p^*(.,j,.)=p(.,j,.)$, and $p^*(.,.,k)=p(.,.,k)$.
In fact, if p* satisfies these constraints then we know that

$$(16) \qquad I(p:\pi)=I(p:p^*)+I(p^*:\pi).$$

Here p* minimizes both $I(.:\pi)$ and $I(p:.)$. The dot indicates the
argument over which the minimization takes place. This demonstrates
that the estimate p* is also the maximum liklihood estimate of p
since $2NI(p:p^*)$ is minus twice the logliklihood ratio. We also
know that p* can be obtained by iterative proportional fitting.
Let us examine the three cases presented above.

1) H_1: $p(i,j,k)=a(i)b(j)c(k)$

Let p(n,i,j,k) denote the n-th iterate in the fitting procedure.

$$p(1,i,j,k)=\{p(i,.,.)/(1/I)\}1/IJK=p(i,.,.)/JK$$

$$p(2,i,j,k)=\{p(.,j,.)/(1/K)\}\{p(i,.,.)/JK\}=p(i,.,.)p(.,j,.)$$
/K

$$p(3.i.j.k)=\{p(.,.,k)/(1/K)\}p(i,.,.)p(.,j,.)/K$$
$$=p(i,.,.)p(.,j,.)p(.,.,k).$$

No other iteration it necessary and p*(i,j,k)=p(i,.,.)p(.,j,.)
p(.,.,k), which are respectively a(i), b(j) and c(k).
(2) H_2:p(i,j,k)=a(i,j)b(j,k)

$$p(1,i,j,k)=p(i,j,.)/K$$

$$p(2,i,j,k)=p(i,j,.)p(.,j,k)/p(.,j,.)$$

$$p(3,i,j,k)=p(2,i,j,k)$$

clearly, the iteration is complete and a(i,j) and b(j,k) can be
determined by examining p(2,i,j,k).
3) H_3:a(i,j)b(j,k)c(i,k) =p(i,j,k)

$$p(1,i,j,k)=p(i,j,.)/k$$

$$p(2,i,j,k)=p(i,j,.)p(.,j,k)/p(.,j,.)$$

$$p(3,i,j,k)=p(i,.,k)/\{\sum_j p(i,j,.)p(.,j,k)/p(.,j,.)\}$$
$$x\ p(i,j,.)p(.,j,k)/p(.,j,.)$$

It would be pleasant if the denominator of the first factor had a
representation as a marginal such as p(i,.,k), but it doesn't
and the iteration does not terminate as the previous two did;
however, the iteration does converge to those functions a(i,j),
b(j,k) and c(i,k). The only drawback is that functions do not have
a simple closed form, but the p* to which the iteration converges
is still a maximum liklihood estimate. The statistic 2NI(p:p*) is
is asymptotically distributed as a χ^2 with the appropriate degrees
of freedom. Note that we write x(i,j,k)=Np(i,j,k); i.e., N=x(.,.,.).
     It should be noted that under H_1 the estimates for x(i,j,k)
are x*(i,j,k)=x(i,.,.)x(.,j,.)x(.,.,k)/N^2 which is the estimate
of x(i,j,k) under the hypothesis of independence of i,j,and k, and
2I(x:x*) is distributed as a χ^2 with IJK-I-J-K+2 degrees of freedom.
     Under H_2, the estimate of x(i,j,k) is x*(i,j,k)=x(i,j,.)x(.,j,k)
/x(.,j,.), which is the estimate of x(i,j,k) under the hypothesis
of conditional independence of i and k given j. 2I(x:x*) is asymptotically distributed as a χ^2 with IJK-IJ-JK+J-1 degrees of freedom.

Finally, under H_3 our estimates do not have closed form, but the hypothesis is equivalent to the hypothesis of 'no second order interaction', and $2I(x:x^*)$ is asymptotically distributed as a χ^2 with $IJK-IJ-IK-JK+I+J+K-1$ degrees of freedom.

Let us consider the following whimsical example as one type of generalization to the problems that we have considered.

Example 2   Professor I.M. Nutty has been doing research for the last 20 years on the long wingedness of certain fruit flies found only in the wilds of southern Ohio and certain suburbs of New York. He injects half of the flies with compound B and the remainder with a placebo. Placebo flies are kept with placebo flies and compound B flies with compound B flies. He then records whether the offspring exhibit long wingedness or not. The offspring were followed for a month and the results summarized in a 2 x 2 contingency table $x(i,j)$ where i=1 indicates a non-placebo offspring, i=2 a placebo offspring, j=1 normal long wingedness, and j=2 the absence of normal long wingedness. The professor wishes to find the estimate of the table under the hypothesis that compound B flies have the same probability of long wingedness as placebo flies. The professor then estimated the table so that it satisfied the following constraints.

$$\begin{pmatrix} 1 & 1 & 1 & 1 \\ 1 & 1 & 0 & 0 \\ 1 & 0 & -1 & 0 \end{pmatrix} \begin{pmatrix} x(1,1) \\ x(1,2) \\ x(2,1) \\ x(2,2) \end{pmatrix} = \begin{pmatrix} 100 \\ 50 \\ 0 \end{pmatrix}$$

The 100 indicates that there were 100 offspring. The 50 indicates that there were 50 non-placebo offspring. The 0 indicates that if hypothesis is correct then $x(1,1)$ and $x(2,1)$ should be equal. It should be noted that the structure of this problem is not so different from the previous except that the last constraint is not a marginal one. Thus we need to expand the theory to cover this situation.

EXTENSION TO ANY SET OF LINEAR CONSTRAINTS

Let us state the problem as follows. Our objective is to minimize $I(p:\pi)$ over the set of all distributions p of finite dimension Ω such that $Cp=\theta^*$ where $\underline{p}'=(p(1),p(2),...,p(\Omega))$, $\underline{\theta}^{*'}=(\theta(1),\theta(2),...,\theta(m))$, and C is an m x Ω matrix with $\Omega \geq m$ and of full rank. C determines the hypothesis we are testing and is called the design matrix. Let us recall that $I(p:\pi)$ is as follows.

$$I(p:\pi)=\sum_{\omega} p(\omega)\ln\{p(\omega)/\pi(\omega)\}$$

Let $C\underline{\pi}=\underline{\theta}$ where $\underline{\pi}'=(\pi(1),\pi(2),...,\pi(\Omega))$. If $\underline{\theta}=\underline{\theta}^*$ then our problem is solved as $I(\pi:\pi)=0$ and this must be the minimum as I is a non-negative function.

Let p* be the distribution that minimizes I and satisfies the constraints $C\underline{p}^*=\underline{\theta}^*$. For the moment, let us assume that $\underline{c}(i)'\underline{\pi} \neq \underline{c}(i)'\underline{p}^*$

for any i where c(i)' is the i-th row of C and c(i,ω) is the (i,ω)-th element of C. Kullback's minimum discrimination information theorem states that

(17) $p*(ω)=\exp\{\Sigma_i \tau(i)c(i,ω)\}\pi(ω)/M(\underline{\tau})$ for some $\underline{\tau}$

where $\underline{\tau}'=(\tau(1),\tau(2),...,\tau(m))$ and is real and M is the moment generating function of the c(i,ω) with respect to π.

(18) $M(\underline{\tau})=\Sigma_ω \exp\{\Sigma_i \tau(i)c(i,ω)\}\pi(ω)$

It should be noted that

(19) $I(p:\pi)\geq\underline{\theta}*\underline{\tau}-\ln\{M(\underline{\tau})\}$

with equality if and only if p=p* and also

(20) $\underline{\theta}*=\nabla_{\underline{\tau}}\ln\{M(\underline{\tau})\}$

where $\nabla_{\underline{\tau}}'=(\partial/\partial\tau(1),\partial/\partial\tau(1),...,\partial/\partial\tau(m))$ is the gradient operator.
    Using these facts, we can construct an iterative procedure that converges to p* in the following manner.
    Expand $\ln\{M(\underline{\tau})\}$ in a multivariate Taylor series about $\underline{\tau}=\underline{0}$. By doing this, we can show that up to terms of the second order

(21) $\underline{\theta}*=\underline{\theta}+\Sigma\underline{\tau}$

where $\Sigma=(\sigma_{ij})$ and $\sigma_{ij}=\Sigma (c(i,ω)-\theta(i))(c(j,ω)-\theta(j))\pi(ω)$. Σ is the covariance matrix of the c(i,ω)'s with respect to π. Now the Taylor expansion of $\ln\{M(\underline{\tau})\}$ at $\underline{\tau}=\underline{0}$ is as follows upto second order terms.

(22) $\ln\{M(\underline{\tau})\}\approx\ln\{M(\underline{0})\}+\{\nabla_{\underline{\tau}}\ln M(\underline{0})\}'\underline{\tau}+\frac{1}{2}\underline{\tau}'\nabla_{\underline{\tau}}'\nabla_{\underline{\tau}}(\ln M(\underline{0}))\underline{\tau}$

One should note the following relations.

(23) $M(\underline{0})=1$. Thus, $\ln M(\underline{0})=0$.

(24) $\nabla_{\underline{\tau}}\ln\{M(\underline{\tau})\}\approx(\Sigma_ω c(1,ω)\exp(\Sigma_i c(i,ω)\tau(i)),...$
        $...,\Sigma_ω c(m,ω)\exp(\Sigma_i c(i,ω)\tau(i))/M(\underline{\tau})$
Thus, $\nabla_{\underline{\tau}}\ln M(\underline{0})=(\theta(1),\theta(2),...,\theta(m))'$.

(25) $\nabla_{\underline{\tau}}'\nabla_{\underline{\tau}}\ln M(\underline{\tau})=(\partial/\partial\tau(i)\{\partial/\partial\tau(j) M(\underline{\tau})/M(\underline{\tau})\})$

Thus we see that (25) evaluated at $\underline{0}$ yields Σ, the covariance matrix of the c's with respect to π. Thus we may rewrite (22) in the following form.

$$(26) \quad \ln\{M(\underline{\tau})\} \approx \underline{\tau}'\underline{\tau} + \tfrac{1}{2}\underline{\tau}'\Sigma\underline{\tau} .$$

By reexamining (24), we can see that

$$(27) \quad \nabla_{\underline{\tau}} \ln\{M(\underline{\tau})\}' = (\Sigma_\omega c(1,\omega)p^*(\omega), \ldots, \Sigma_\omega c(m,\omega)p^*(\omega))$$

as $p^*(\omega) = \exp(\Sigma c(i,\omega)\tau(i))\pi(\omega)/M(\underline{\tau})$ by Kullback's minimum discrimination information theorem. Thus, it is evident that

$$(28) \quad \nabla_{\underline{\tau}} \ln\{M(\underline{\tau})\}' = \underline{\theta}^*.$$

Thus, we know many things about the 'true' $\underline{\tau}$. Let us use these facts. By rewriting (28) with $\ln M(\underline{\tau})$ replaced by the right hand side of (26), we obtain $\underline{\theta}^* = \nabla_{\underline{\tau}}\{\underline{\theta}'\underline{\tau} + \tfrac{1}{2}\underline{\tau}'\Sigma\underline{\tau}\}$ which yields:

$$(29) \quad \underline{\theta}^* = \underline{\theta} + \Sigma\underline{\tau}.$$

We know from (19) that $I(p^*:\pi) = \underline{\theta}^*\underline{\tau} - \ln M(\underline{\tau}) = \underline{\theta}^*'\underline{\tau} - \underline{\theta}'\underline{\tau} - \underline{\tau}'\Sigma\underline{\tau}$ which is gotten by substituting the right hand side of (26) for $\ln M(\underline{\tau})$. As C is of full rank we know that Σ is invertible. Thus, we get $I(p^*:\pi) \approx \tfrac{1}{2}(\underline{\theta}^* - \underline{\theta})'\Sigma^{-1}(\underline{\theta}^* - \underline{\theta}) \approx \tfrac{1}{2}\underline{\tau}'\Sigma\underline{\tau}$. Let $f(\underline{\tau}) = \tfrac{1}{2}\underline{\tau}'\Sigma\underline{\tau}$. By finding the $\underline{\tau}$ for which $\nabla f(\underline{\tau}) = \underline{0}$ we will have found the $\underline{\tau}$ for which $I(p^*:\pi)$ attains a local minimum which in this case is the global minimum as I is a convex function of $\underline{\tau}$.

Let us expand $f(\underline{\tau})$ about an arbitrary point, say $\underline{\tau}_k$. This yields the following upto to second order terms.

$$(30) \quad f(\underline{\tau}) = f(\underline{\tau}_k) + (\nabla_{\underline{\tau}} f(\underline{\tau}_k))'(\underline{\tau} - \underline{\tau}_k)' + \tfrac{1}{2}(\underline{\tau} - \underline{\tau}_k)'\nabla_{\underline{\tau}}'\nabla_{\underline{\tau}} f(\underline{\tau}_k)(\underline{\tau} - \underline{\tau}_k).$$

By taking the gradient of (30), we obtain

$$(31) \quad \nabla_{\underline{\tau}} f(\underline{\tau}) = \nabla_{\underline{\tau}} f(\underline{\tau}_k) + \nabla_{\underline{\tau}}'\nabla_{\underline{\tau}} f(\underline{\tau}_k)(\underline{\tau} - \underline{\tau}_k).$$

Denote $\nabla_{\underline{\tau}}'\nabla_{\underline{\tau}} f(\underline{\tau}_k)$ by S_k and note that it is the Hessian of f at $\underline{\tau}_k$. If we equate (31) to $\underline{0}$ and solve of $\underline{\tau}$ we have the ordinary Newton-Raphson iteration; which, if we denote $\underline{\tau}$ by $\underline{\tau}_{k+1}$ yields:

$$(32) \quad \underline{\tau}_{k+1} = \underline{\tau}_k - S_k^{-1}\nabla_{\underline{\tau}} f(\underline{\tau}_k)$$

where S_k^{-1} is the inverse of the Hessian of f at $\underline{\tau}_k$. It is well known that this iteration always converges provided the S_k are positive definite, which they are. Noting that $\underline{\theta}^* - \underline{\theta}_k = S_k\underline{\tau}_k$ and $\nabla_{\underline{\tau}} f(\underline{\tau}_k) = S_k\underline{\tau}_k$, we are able to conclude

$$(33) \quad \underline{\tau}_{k+1} = \underline{\tau}_k - S_k^{-1}(\underline{\theta}^* - \underline{\theta}_k).$$

The sequence τ_k converges to the minimizing τ, and as a fortuitous by-product S_k^{-1} converges to Σ^{-1}, the covariance matrix of the estimates of the τ's; i.e., the Hessian of f at the minimizing τ.

Now suppose that we relax the condition that $\underline{c}(i)'\underline{p} \neq \underline{c}(i)'\underline{\pi}$ for any i. Let $\underline{c}(i)'\underline{p} = \underline{c}(i)'\underline{\pi}$ for $i=1,2,\ldots,m_1$ and $\underline{c}(i)'\underline{p} \neq \underline{c}(i)'\underline{\pi}$ for $i=m_1+1, m_1+2, \ldots, m$. Let $\underline{\theta}*' = (\underline{\theta}_1*', \underline{\theta}_2*')$ where $\underline{\theta}_1$ is an m_1 x 1 matrix and $\underline{\theta}_2$ is an m_2 x 1 matrix with $m_2 = m - m_1$. Similarly, we partition the matrix C into two matrices C_1 and C_2. C_1 consists of the first m_1 rows of C and C_2 the remaining m_2 rows of C. A similar partition of τ allows us to rewrite the relation $\underline{\theta}* = \underline{\theta} + \Sigma\underline{\tau}$ as

$$(34) \quad \begin{pmatrix} \underline{\theta}_1* \\ \underline{\theta}_2* \end{pmatrix} = \begin{pmatrix} \underline{\theta}_1* \\ \underline{\theta}_2* \end{pmatrix} + \begin{pmatrix} \Sigma_{11} & \Sigma_{12} \\ \Sigma_{21} & \Sigma_{22} \end{pmatrix} \begin{pmatrix} \underline{\tau}_1 \\ \underline{\tau}_2 \end{pmatrix}$$

where $\Sigma = (\Sigma_{ij})$ and Σ_{ij} is m_i x m_j. (34) can be rewritten as the following two equations.

$$(35) \quad \underline{\theta}_1* = \underline{\theta}_1 + \Sigma_{11}\underline{\tau}_1 + \Sigma_{12}\underline{\tau}_2$$

$$\underline{\theta}_2* = \underline{\theta}_2 + \Sigma_{21}\underline{\tau}_1 + \Sigma_{22}\underline{\tau}_2$$

The conditions that $\underline{c}(i)'\underline{p} = \underline{c}(i)'\underline{\pi}$ for $i=1,2,\ldots,m_1$ imply that $\underline{\theta}_1* = \underline{\theta}_1$. Hence, we conclude that $\Sigma_{11}\underline{\tau}_1 = -\Sigma_{12}\underline{\tau}_2$, which implies that $\underline{\tau}_1 = -\Sigma_{11}^{-1}\Sigma_{12}\underline{\tau}_2$. Thus, we may write $(\underline{\tau}_1, \underline{\tau}_2) = \underline{\tau}_2'(-\Sigma_{21}\Sigma_{11}^{-1}, I_{m_2})$ where I_{m_2} is the identity matrix of order m_2. Using this last identity in the quadratic representation of $I(p*:\pi)$ we can show that:

$$(36) \quad I(p*:\pi) = \tfrac{1}{2}\underline{\tau}_2'\Sigma_{22.1}\underline{\tau}_2$$

where $\Sigma_{22.1} = \Sigma_{22} - \Sigma_{21}\Sigma_{11}^{-1}\Sigma_{12}$. Under the relaxed conditions the iteration is exactly the same as before except that at each step S_k is replaced by the estimate of $S_{22.1}$ which is gotten by partitioning S_k and forming the required quantity.

As a concluding remark, it should be noted that much interest has been aroused in contingency table analysis in the past several years. The advent of the computer has made many algorithms available to researchers and practising statisticians. The two algorithms discussed in this paper are available as part of a package of programs for contingency table analysis and are distributed at a nominal cost from the statistics department of the George Washington University, Washington, D.C.

References
(1) Gokhale, D.V. and Kullback, S.(1978). The Information in Contingency Tables. Marcel Dekker, Inc.
(2) Ireland, C.T. and Kullback, S. (1968). Contingency Tables with Given Marginals, Biometrics 24, 707-713.

CONTAB*

John Nolan

The George Washington University
Washington, D.C., USA

1. INTRODUCTION

The CONTAB program, written in PL/1, is available on magnetic tape files which are being distributed at cost by the Department of Statistics of The George Washington University (Washington, D.C., 20016). Each file consists of a sequence of 80-column card images containing EDCDIC coding of PL/1 Optimizer Compiler statements. Some changes are necessary for use with the F Compiler.

In the following sections, FORTRAN symbols for multiplication (*), division (/), and exponentiation (**) are used.

CONTAB computes maximum likelihood estimates of parameters in the log-linear model for contingency

---

* This paper is taken from a recent unpublished report authored jointly by Marian Fisher, C. Terrence Ireland, John Keegal, Solomon Kullback, John Nolan and Frederick Scheuren, all of whom are (or have been) associated with the Department of Statistics of The George Washington University, Washington, D.C. 20016. The document is titled Computer Programs for Contingency Table Analysis. The principal author of this particular section is John Nolan. Certain editorial changes have been made by the editor to meet the objectives of these Proceedings.

tables and computes the associated test statistics.
It prints a number of statistics, for example, standardized effects and outlier statistics, useful for practical
data analysis.

RUNNING THE PROGRAM

Once the program is catalogued on a program library,
it can be invoked using a deck setup similar to the
following:

```
// jobname JOB etc.
// stepname EXEC PGM=contab, REGION=128K
// STEBLIB DD DSN=library, DISP=SHR
// SYSPRINT DD SYSOUT=A
// PRINT DO SYSOUT=A
// SYSIN DD *

            - - - - - - - - - - -
            table parameters & data
            - - - - - - - - - - -

// LABELS DD *

            - - - - - - - - - - -
            level labels (optional)
            - - - - - - - - - - -

//
```

Since level labels are optional, the DD card and
data need only be included if the "LEVELS" flag is set
(see input parameters).

The initial core allocation of 128K will serve for
most small tables. If the program terminates with an
80A ABEND, the allocation should be increased.

INPUT KEYWORDS + DATA

It is possible to analyze several tables with one
run and to test more than one hypothesis against each
table. The input is divided into table parameters,
table data, hypothesis parameters, and hypothesis data.
Each table and hypothesis is defined by keywords. These
keywords have default values which remain in effect
until changed.

When placing keywords on input cards, keywords may
be specified in any order and are followed by values

("KEYWORD = VALUE"). Keywords and values may be
placed anywhere on the card, with leading and trailing
blanks allowed. More than one keyword may be placed
on a card at the user's convenience. Numeric values
may be placed on a card at the user's convenience.
Numeric values may be specified as integers, floating
point decimal values with decimal point, or floating
point decimal values with exponent. String values are
specified as any characters between single quotes (').
Boolean values are either '1'B or '0'B. A leading
minus sign "-" is used to indicate negative numeric
values.

Examples of properly formed input are:

KEYWORD = 3 KEYWORD2 = 7.654 STRINGA = 'HI
HARRY'

KEYWORD5 = .34E-02 KEYWORD4 = .66E03
BOOLEANX = '1'B

KEYWORD3 = .5 KEYWORD6 = 0.5

The order of input is: Table keywords and their
values followed by a single semi-colon ";", then table
data values, then hypothesis keywords and their values
followed by a single semi-colon then hypothesis data
values, then either another set of hypothesis keywords
and values or a new set of table keywords and values.

TABLE KEYWORDS

FACTORS
This mandatory keyword specified the
number of dimensions in the table. The
integer value must be less than 20.
Each factor represents a characteristic
used to describe the data.

LIST
This option specified which output
listings are to be printed. LIST
applies to both tables and hypothesis.
LIST is specified as a string containing
one or more of the letters from the set
'NDRMEO'. 'R', 'M', 'E', and 'O' apply
to hypotheses and will be described in
the section on hypothesis keywords. 'N'
simply cancels all other options. If
you specify LIST = 'N' along, all optional
listings are turned off. If 'N' is the
first letter of a LIST string, only

listings requested in the remainder of
the string are in effect, all others are
canceled. 'D' requests a listing of the
original table data. Each element is
listed modified with either "SMOOTH" or
"ZERO". Missing entries are given a
list value of ZERO and underlined. The
default value for LIST is LIST = 'N'.

ERRORMAX ERRORMAX specified the convergence
criterion for hypothesis iteration.
The default value is ERRORMAX = .001.

ZERO This value is added to all cells containing true 0.00. Note that if SMOOTH is
non-zero, ZERO is not added to cells.
Default is .000001.

TITLE For table identification, a string of
less than 50 characters will be printed
at the top of each page of the output.
Since the keyword is a string keyword,
values are of the form TITLE = 'ANY
STRING'. If no title is given, none is
printed. The title may be changed as
needed.

FL(k) This keyword provides the means to assign
labels to factors. The assignment is of
the form: FL(k) = 'A STRING' where k is
the number of the factor and the value is
a string less than 12 characters long.
FL(1) assigns the value string to factor
1, FL(2) to factor 2 and so on. Since
the number of factors is limited to 20,
the number of label assignments is
limited to 20. Default labels are the
letters "A", "B", "C", etc. for factors
1, 2, and 3, etc. respectively.

OUTBOUND The optional numeric keyword supplied a
threshold for outlier listing. If
LIST = 'O' (letter 'O') is specified,
then outlier values greater than OUTBOUND
will be listed in the hypothesis summary.
This listing consists of subscript values
of the cell and its outlier value. The
default for OUTBOUND is 7.0.

MAXINT The integer MAXINT sets an upper bound

on the number of factors involved in the calculation of effects. For example, MAXINT = 2 allows only the main effects and the first-order interactions (those involving 2 factors). The default value is MAXINT = number of factors.

SMOOTH   The value of SMOOTH is added to each cell of the contingency table (except missing values represented by a negative number). SMOOTH may be changed for each hypothesis. The default is 0. The cell count + SMOOTH becomes the table used for all subsequent calculations.

LEVELS   This boolean keyword identifies tables in which the individual levels of each factor are identified. The actual labels are supplied in the LABELS input file. The number of labels required is equal to the sum of the number of levels of all factors. For example, a 4x3x2 table needs 4+3+2=9 level labels. Use of level labels does not preclude use of factor labels (see FL keyword).

TABLE DATA

Data for each table consists of 2 parts: a) a list of the number of levels of each factor and b) the cell counts of the table. The factor list consists of a list of integers which define for each factor the number of levels (number of subscripts) in that factor. The first integer defines the first factor (factor 1), the second integer defines factor 2 and so on. A table can be redefined by changing the factor list without changing the actual cell counts. For example, a 4x6 table (defined by the list 4 6) can be redefined as a 4x3x2 table (defined by the list 4 3 2) without changing the cell counts (both tables require 24 counts). The 4x6 table is a 2-dimensional table described by two factors containing a total of (4+6) = 10 levels. The 4x3x2 table is 3-dimensional (3 factors) with (4+3+2) = 9 levels.

After the factor list, the cell counts for the table follow. There is no separator between the factor list and the cell counts. It is important to realize that the value given for the keyword FACTORS determines the length of the factor list. If the value for FACTORS is 3 then the first 3 values will be taken to be the

factor list, the remaining values will be assigned as
cell counts. The factor list is really a subscript
list, not table data.

    After the factor list, the cell counts are given
so that the last subscript varies most rapidly and the
first subscript varies least rapidly. For example, the
simple 2x3 table shown here:

                        FACTOR 2

	23	45	67
FACTOR 1	89	90	12

would be input: 2, 3, 23, 45, 67, 89, 90, 12
Of course, 2, 3 is the factor list and the rest of the
integers are the cell counts from the first row then
from the second row. Notice that the first row counts
are exhausted before any counts are taken from the
second row. For a 4x3x2 table (3-dimensional) the cell
counts are taken in this subscript order:

    (1,1,1), (1,1,2) (1,2,1) (1,2,2), ..., (4,3,1),
(4,3,2)

    Although the values shown above are separated by
commas, spaces are sufficient to separate values. Two
sequential commas ",," <u>will</u> result in erroneous input.

HYPOTHESIS KEYWORDS

TERMS   This mandatory keyword specifies the
                    number of marginals held fixed under the
                    hypothesis. TERMS is used to get the
                    length of the marginal factor list
                    described under HYPOTHESIS DATA below.
                    The lowest value possible for TERMS is 1.

BASE   This boolean keyword identifies the
                    hypothesis on which I* is based for all
                    hypotheses following in the same table.
                    I* is the percent difference of the
                    Information Statistic for a hypothesis
                    from the base hypothesis. The only
                    permitted value is BASE = '1'B. BASE
                    <u>must</u> be respecified for each new table.
                    <u>BASE</u> can be respecified at any time.

SMOOTH   As noted under TABLE KEYWORDS above, the
                  value for SMOOTH can be changed for each
                  hypothesis.

LIST   For hypotheses, the LIST letters 'R',
                  'M', 'E', and 'O' apply. 'R' specifies
                  Residuals; 'M' specifies Marginals; 'E'
                  specifies Effects; 'O' specifies Outliers.
                  'O' implies 'R'. For details of listings,
                  see OUTPUT below. See also the explana-
                  tion of LIST in TABLE KEYWORDS. See the
                  explanation of OUTBOUND in TABLE KEYWORDS
                  if you include LIST = 'O' since that
                  keyword will probably require a new
                  (non-default) value.

HYPOTHESIS DATA

     After hypothesis keywords for a hypothesis,
hypothesis data follows in two parts: a) the marginal
factor list and b) factor numbers. The length of the
marginal factor list is equal to the value of TERMS.
Each value in the list is the number of factors in
each marginal set held fixed under the hypothesis.
After the marginal factor list, the factor numbers of
the factors held fixed are given. The length of the
list of factor numbers is equal to the sum of the
integers in the marginal factor list. Each integer
in the marginal factor list specifies the number of
values in the associated factor list. Figure
shows the relationship between TERMS, the marginal
factor list, and the factor numbers

FIGURE
Hypothesis Data Stream Relationships

TERMS

factor	factor		factor	
count	count		count	..... etc.
1	2		3	

- - - - - - - - - - - - - - - number of items
in this marginal
held fixed

factor	factor	factor	factor	factor	factor	factor	factor
#	#	#	#	#	#	#	#

HYPOTHESIS DATA EXAMPLES

For each example, only the value of the TERMS
keyword is given (all other hypothesis keywords are
omitted). Following the required semi-colon, the
marginal factor list and the factors are shown as they
are required in the input stream. All examples shown
are for a 3 factor table. For degrees of freedom
calculations use I for the number of levels in the first
factor, J for the number of levels in the second factor
and K for the number of levels in the third factor.
The tables considered would be therefore IxJxK in size.

1 Three way independence:

TERMS = 3; 1,1,1, 1, 2, 3

This hypothesis implies that each factor is independent
of all other factors. Each marginal set involves 1
factor. Marginal sets not fixed are: (1,2), (1,3),
(2,3), and 1,2,3). There are

$$((I-1)*(J-1))+((I-1)*(K-1))+((J-1)*(K-1))+((I-1)*(J-1)*(K-1))$$

degrees of freedom.

    2 Independence of the third factor from the first two:

        TERMS = 2; 2, 1, 1, 2, 3

This hypothesis involves 2 marginal sets, the first with 2 factors and the second with one. Marginal sets not fixed are: (1,3), (2, 3) and (1,2,3). There are:

$$((I-1)*(K-1))+((J-1)*(K-1))+((I-1)*(J-1)*(K-1))$$

degrees of freedom.

    3 Conditional independence of the first 2 factors given the third:

        TERMS = 2; 2, 2, 1, 3, ?, 3

Unfixed sets are (1,2) and 1,2,3). Degrees of freedom are

$$((I-1)*(J-1))+((I-1)*(J-1)*(K-1)).$$

    4 No second order interaction:

        TERMS = 3; 2, 2, 2, 1, 2, 1, 3, 2, 3

This hypothesis allows no second order interaction. Unfixed set is (1,2,3). There are $((I-1)*(J-1)*(K-1))$ degrees of freedom.

    5 All marginals held fixed:

        TERMS = 1; 3, 1, 2, 3

The set (1,2,3) implies that the entire table is required, i.e., no simpler model than the table is possible to represent the table. This model is, of course, a perfect fit to the table (itself). This hypothesis will obtain a full set of effects (but be sure to include LIST = 'E'). For a 4 factor table, the required input would be: TERMS = 1; 4, 1, 2, 3, 4. Degrees of freedom and information statistic calculations for this hypothesis are not meaningful.

OUTPUT

    For each table, an output listing is produced
containing at <u>least</u> the necessary data on each hypothesis
and any other listing the user requires. Each hypothesis
is numbered consecutively from 1 without regard for new
tables. Additionally, a summary of all hypotheses tried
is printed at the end of the regular listing. This
summary contains some of the information for the hypothesis plus the outlier information if that is requested.
The listings produced are in this order:

LISTING	OPTIONAL	CONTENTS
DATA	YES	Value for each cell and a log ratio (LIST = 'D')
MARGINALS	YES	Marginal totals under this hypothesis (LIST = 'M')
RESIDUALS	YES	Predicted, residual, outlier values, and log ratios for the predicted values (LIST = 'R')
EFFECTS	YES	Interactions estimated under the hypothesis (LIST = 'E')
HYPOTHESIS	NO	Result of applying the hypothesis to the table.
SUMMARY	NO	A summary of the hypotheses for all tables. Outlier listing is given if LIST = 'O'.

A detailed explanation of each output list follows.

TABLE DATA

    If LIST = 'D' is specified, a listing of the input
contingency table is printed. This will be helpful in
checking the input. The table is printed as a group of
sub-matrices in which all subscripts except the last 2
are held fixed. Thus a JxK table is printed directly;
a IxJxK table is printed as a series of JxK matrices
while i takes on successive values 1, 2, 3, ..., I-1,
I. The TITLE, if any, appears at the top of the page.
Each submatrix is listed with an indication of the
higher order subscripts which are held fixed.

    Within each submatrix, the entries for each cell
are as follows:

    <u>OBSERVED</u> This value is the original cell value
modified by the value of SMOOTH. If the cell value is
0.0, and SMOOTH is 0.0, then the cell value is set to

ZERO (a small value nominally .000001). Any entry
which is missing is underlined. The value is shown
as 0.0. Missing cells are those represented as
negative values on input.

LOG-RATIO This value is the log (ln or base e)
of the ratio of the observed value for the cell to
the observed value of the cell with the highest subscript used as a reference cell. For example, in a
4x3x2 table, the reference cell is cell (4,3,2). The
choice of this cell is arbitrary and does not affect
calculations.

MARGINALS

If LIST = 'M' is specified, marginal totals are
listed for this hypothesis. If the hypothesis is
that all the factors are independent, the number of
marginal totals is equal to the sum of the number of
levels of all the factors. If, however, the hypothesis
is not independence, some marginal totals are broken
down into subtotals. This increases the number of
marginal totals for the hypothesis.

Any factor that is independent has as many
marginal totals as there are levels in the factor.
When two or more factors are combined into a marginal
set their marginal totals as independent factors are
deleted and new marginal totals are created. The
number of marginal totals created is equal to the
product of the number of levels in each of the factors.

For example, consider the viewing habits (from
1 to 4 hours per week) of television viewers (3 age
groups) using 2 types of television sets. There would
be 2+4+3=9 marginal totals in a contingency table
(under independence.) If we combine factors 1 (Type Of
Set) and 3 (Age) into a marginal set, we will have 4
marginal totals for factor 2 and 6 marginal totals
(2*3) for factors 1 and 3 combined. Table   shows
the marginal totals and the cells added to get them.

FIGURE
Television Contingency Table Marginal Totals

MARGINAL TOTAL	CELL MEMBERSHIP
1, 1	a, d, g, j
1, 2	b, e, h, k
1, 3	c, f, i, l
2, 1	m, p, s, v
2, 2	n, q, t, w
2, 3	o, r, u, x

HOURS PER WEEK VIEWING

1	a, b, c, m, n, o
2	d, e, f, p, q, r
3	g, h, k, s, t, u
4	j, k, l, v, w, x

If we rearranged the marginal sets so that Factors 1 &
3 are in one set and Factors 2 * 3 are in another, we
would get (2 * 3) + (4 * 3) = 18 marginal toals shown
in Figure

FIGURE
Television Contingency Table Marginal Totals

MARGINAL TOTAL	CELL MEMBERSHIP
SET TYPE * AGE GROUP	
1, 1	a, d, g, j,
1, 2	b, e, h, k
1, 3	c, f, i, l
2, 1	m, p, s, v
2, 2	n, q, t, w
2, 3	o, r, u, x
SET TYPE * HOURS PER WEEK VIEWING	
1, 1	a, m
1, 2	b, n
1, 3	c, o
2, 1	d, p
2, 2	e, q
2, 3	f, r
3, 1	g, s
3, 2	h, t
3, 3	i, u
4, 1	j, v
4, 2	k, w
4, 3	l, x

RESIDUALS

If LIST = 'R', or LIST = 'O' (implying 'R') then residual and outlier values are listed. The residual table is printed as a group of sub-matrices in which all subscripts except the last two are held fixed. The subscripts held fixed are printed at the heading of each sub-matrix.

In each sub-matrix, the entries for each cell are as follows:

OBSERVED   This is the original cell value modified by either SMOOTH or ZERO.

PREDICTED   This is the estimated value for the cell under the hypothesis. If the hypothesis is independence of all factors, the estimated value is the product of all the marginal toals which include the cell, divided by (SAMPLE SIZE) ** (FACTORS - 1) where ** denotes exponentiation. If the hypothesis is more complex than simple independence, computation of the PREDICTED value may be an iterative procedure.

RESIDUAL   This is OBSERVED-PREDICTED for the cell

OUTLIER   This value is the lower bound of the decrease in the Information Statistic that would occur if the cell were deleted from the model. The OUTLIER value is: $2*V*\ln(V/P)+(N-V)*\ln((N-V)/N-P))$ where V = OBSERVED, P = PREDICTED, N = SAMPLE SIZE. The OUTLIER value is a measure of the "oddness" of the observed value; cells with higher OUTLIER values fit the model less well than those with lower values. Too many high outlier values may indicate an inappropriate or poorly fit model.

LOG-RATIO   This ratio is the log (base e) of the PREDICTED value for the cell to the PREDICTED value for the reference cell (cell with the highest subscript value).

EFFECTS

If LIST = 'E' is specified, a list of effects is printed. The number of effects is controlled by the keyword MAXINT and by the hypothesis itself. Effects are listed in order of increasing number of factors. For each hypothesis, the GENERAL MEAN is listed which is:

$$\frac{\text{sum of }(\ln(P) - \ln(N))}{\# \text{ of cells}}$$

where P = PREDICTED and N = SAMPLE SIZE.

Then each set of effects is listed as a group of sub-matrices where all but the last two subscripts are held fixed. For each sub-matrix, the heading lists the factor labels of the effects and the subscripts which are held fixed.

The following items are listed:

EFFECT The computed effect of the given factor or combination of factors on this cell.

STAN.DEV. The standard deviation for the effect

STANDARDIZED The ratio EFFECT/STAN.DEV. The STANDARDIZED value is extremely useful in forming the model since interactions are more statistically noticeable for this row than the effects row.

HYPOTHESIS

For each hypothesis, a short listing represents the result of applying the hypothesis to the contingency table. The hypothesis number is shown in the heading. Hypotheses are numbered consecutively from 1 without regard to the number of tables. For each hypothesis, the following information is given:

INFORMATION STATISTIC The information statistic is 2 * sum of OBSERVED * ln(OBSERVED/PREDICTED). Please note that the (different) chi-squared statistic is the sum of $(\text{OBSERVED-PREDICTED})^2$/PREDICTED.

PROBABILITY OF A GREATER VALUE ·This is the significance level of the information statistic when compared with the appropriate chi-squared distribution. It is a measure of the confidence that you can have in accepting or rejecting an hypothesis. A probability of -1 is a flag indicating that computation of this probability is meaningless.

LN(REFERENCE/(N/NUMBER OF CELLS)) Ln of the cell value of the cell with the highest subscript value (the reference) divided by the average count per cell.

SMOOTH   Value of the keyword SMOOTH

ZERO   Value of the keyword ZERO

SAMPLE SIZE   Total number of observations in the table.   The sum of all the cells (unsmoothed).

MAXERROR   Value of the convergence threshold.

DEGREES OF FREEDOM   This is the number of independent zero parameters under the hypothesis.

FACTORS   The number of factors (dimensions) in the table.

NUMBER OF CELLS   Total number of cells in the table. This is the product of the number of levels of all the factors.

ZERO CELLS   The number of cells where the cell count is zero (unsmoothed).

MARGINAL ZERO TOTALS   Number of marginal totals which are zero (note that this depends on the hypothesis)

I*   The percent difference of the Information Statistic for this hypothesis compared to the BASE hypothesis.   This appears only when a BASE hypothesis is specified.

CORRECTED I*   The value of I* corrected for degrees of freedom.   CORRECTED I* = 1.0 - DFB * (1.0 - I*)/DF where DFB is the number of degrees of freedom for the BASE hypothesis, DF is the number of degrees of freedom for this hypothesis, and I* is the uncorrected percent difference.

MISSING OBSERVATIONS   The number of observations whose value was missing (represented by a negative number) on input.   If none are missing, this entry will not appear.

OUTLIER BOUND   The value of the outlier threshold OUTBOUND.   Only appears if LIST = '0' is specified.

NUMBER OF OUTLIERS   The number of outlier values which exceed the threshold OUTBOUND.   Only appears if LIST = '0' is specified.

EFFECTS - INTERACTION LEVEL PRINTED   The level
of interactions printed. Appears only if LIST = 'E'
is specified.

MARGINALS   Flag word appears if LIST = 'M' is
specified.

RESIDUALS   Flag word appears if LIST = 'R' is
specified.

NESTED ON HYPOTHESIS   The hypothesis number of
the BASE hypothesis.

DIFFERENCE INFORMATION STATISTIC   The difference
between the information statistic for this hypothesis
and the BASE hypothesis. Appears only if a BASE
hypothesis is specified.

PROBABILITY OF A GREATER VALUE   Significance level
of the information statistic for this hypothesis to
the information statistic for the BASE hypothesis.
Appears only if a BASE hypothesis is specified.

DIFFERENCE DEGREES OF FREEDOM   The difference
between the degrees of freedom for this hypothesis
and the BASE hypothesis. Appears only if a BASE
hypothesis is specified.

HYPOTHESIS SUMMARY

After all listings for tables and hypotheses, a
summary is printed of all hypotheses in condensed
form, the information below is listed left to right.

HYPOTHESIS NUMBER   The number of the hypothesis.
Hypotheses are numbered consecutively from 1. Use
TITLE to keep track of which table the summary information
applied to. To the right of the hypothesis number, the
listings requested for this hypothesis ("MARGINALS",
"EFFECTS", or "RESIDUALS") are shown. Below the listings entry the factor labels of the marginal sets held
fixed are shown. Each line represents a marginal set.
The number of lines is equal to the value of TERMS.

OUTLIER   If an outlier listing is requested, and
if any outlier values exceed OUTBOUND, the subscript
numbers of the cell or level names and the value of
OUTLIER are shown.

SMOOTH   Value of the keyword SMOOTH.

ZERO   Value of the keyword ZERO.

I.S.   The Information Statistic for this hypothesis.

D.F.   Number of degrees of freedom for this hypothesis.

PROB   Probability of a greater value.   The significance level for the Information Statistic.

I*   The percent difference of the Information Statistic for this hypothesis compared to the BASE hypothesis.   Shown as 0.00 if no BASE hypothesis is specified.

CI*   The value of I* corrected for degrees of freedom.   Shown as 0.00 if no BASE hypothesis is specified.

TABLE

Keyword Summary

KEYWORD	OPTIONAL	USE	DEFAULT	RANGE OF VALUES
BASE	YES	HYPOTHESIS	--	'1'B
ERRORMAX	YES	TABLE	.001	ERRORMAX > 0
FACTORS	NO	TABLE	--	2 < FACTORS < 20
FL(k)	YES	TABLE	"A" - "T"	k < FACTORS; length < 12 characters
LEVELS	YES	TABLE	--	'1'B
LIST	YES	BOTH	'N'	'NDRMEO'
MAXINT	YES	TABLE	FACTORS	1 < MAXINT < FACTORS
OUTBOUND	YES	HYPOTHESIS	7.0	
TERMS	NO	HYPOTHESIS	--	1 < TERMS
TITLE	YES	TABLE	' '	length < 50 characters
ZERO	YES	TABLE	.000001	0 < ZERO

PART II

APPLICATIONS

THREE STATISTICAL TREATMENTS OF CONTINGENCY TABLES IN ROAD SAFETY STUDIES

S. LASSARRE

Organisme National de Sécurité Routière
Arcueil, France

ABSTRACT. Three exemples covering important research fields in road safety are presented after 1. Introduction : 2. Effectiveness evaluation, 3. Risk study and 4. Epidemiological study.

The statistical techniques come from inferential statistics (experiment and test theory) and from data analysis (segmentation, correspondence analysis, clustering).

1. INTRODUCTION

A lot of contingency tables are computed in road safety field, specially within the administrative statistics forms, but most of them remain ignored and unanalyzed. On the other hand, for special problems where the ultimate object is not the description of the phenomena in itself, but rather a reliable conclusion on which we can base a decision, the contingency table analysis techniques are used more frequently.

I want to introduce to you three case-studies achieved within O.N.S.E.R. Evaluation Center activities, where a statistical treatment of contingency table has been carried out :

- Effectiveness evaluation of a countermeasure with experimental design and statistical test on 2 x 2 table.

- Risk study with segmentation of a drivers population involved in an accident according to various criteria.

- Epidemiological study with correspondence analysis and clustering of French departments according to accident characteristics for a purpose of primary safety survey preparation.

2. EFFECTIVENESS EVALUATION OF A COUNTERMEASURE

2.1. Design and test problem in evaluation

Such a study always begins by designing either a true experiment, if statistical units, on which the observation is done, are random assigned to the control or the experimental ; or a quasi-experiment, if the experimental is choosen in a first time, the control in a second time with matching criterion.

We can illustrate in diagram these designs from using these standard symbols [5] [15] :

X = exposure to project activity or treatment
O = process of observation or measurement
R = random assignment of primary statistical units (individuals or
  road sections) to experimental or control
r = random assignment of secondary statistical units (people groups,
  area, roads) to experimental or control.

The first one is a strong design either with randomization on primary units :

```
        Experimental     O    X    O
    R
        Control          O         O
```

or with randomization on secondary units plus a careful pre-experimental matching

```
        Experimental     O    X    O
    r
        Control          O         O
```

It is very strong if replication with a rank ordering on pairs of secondary units is carried out.

The second one is a weak design with no randomization and only an a posteriori matching.

```
    Experimental     O    X    O

    Control          O         O
```

This design is subject to biais and specially to the regression to the mean if the experimental is choosen among the extreme.

The two observation periods before and after treatment can have different lenght but must cover the same months in view to elimi-nate seasonnality.

The collected data are the number of accidents described by some parameters in interaction with the countermeasure (gravity, acci-dent type). The experiment outputs are summarized in a 2 x 2 con-tingency table [3], [9]

	Before	After
Control	τ_1	τ_2
Experimental	τ_3	τ_4

where are presented accident rate or count according to the gravity or accident type. We use as evaluation criterion the per-centage reduction of accident rate expressed by the formula

$$ev = \frac{\tau_3 \times \frac{\tau_2}{\tau_1} - \tau_4}{\tau_3 \times \tau_2/\tau_1} = 1 - \frac{\tau_4/\tau_3}{\tau_2/\tau_1}$$

where $\tau_3 \times \tau_2/\tau_1$ expressed the expected accident rate on the experimental.

We use the chi square test on the number of accident to judge the significativity of the reduction. We generaly end the study by an economic evaluation of project rentability.

2.2. Case-studies

Several studies have been carried out at the Evaluation Center regarding effectiveness of countermeasures : guardrails on central reserve motorways [7], gard-rails before trees [8], alarm-speed device [16], horizontal lateral-side and median marking road [10], delineators [11], concrete groowing of motorways pavement [12]. I will develop these two last cases.

For delineator study [11], 180 kilometers of experimental sections have been choosen on a main road into four departments, and mat-ched with 200 kilometers of control sections of three main roads on criteria of traffic flow and physical road characteristic. The before period last two years, the after period one year. We came to the following conclusions :

. 9 % reduction of accidents rate (non significant)
. the effectiveness is near zero in day time and without adverse
  weather conditions
. on the contrary, 19 % reduction of accidents by night (non significant) and 31 % reduction with adverse weather (significant).

The delineatorsare specially effective guidence devices by night
and with adverse weather. A second year of observation has shown
that the night effectiveness became near zero and the adverse
weather conditions effectiveness had diminished by half. The
reasons of such a change have not been analysed. But this stresses
the need and utility of a project following up in view to define
its life.

For the concrete groowing study [12], we dispose of 9 kilometers
long experimental sections on an urban motorway in Paris South
suburb. They are matched with 28 kilometers long controlled sections of the same motorway according to slide rate and traffic
flow. The before and after period last one year. We infer from
this study that the accident rate reduction due to groowing is
equal to 45 % without differenciation on wet or dry surface.

2.3. Conclusion

In parallel with these evaluation studies, we begin developping
an evaluation methodology concerning experimental design and proper
statistical tests [3], [13]. However, we can notice these last
years a slowing down of evaluation studies. Many experimental
designs suffer from various threats to validity and don't give an
effectiveness evaluation with sufficient internal and external
validity. So, from our and others countries evaluation researchs,
we have to make an effort to achieve a complete evaluation methodology to gain the decision maker trust.

3. RISK STUDY

3.1. Introduction

Risk study [4], [6] in order to clarify the target choice in a
decision process, aims to detect sub-populations of vehicle-drivers
having anormally high risk. The risk is defined as an accident frequency divided by an exposure or occurrence frequency on road.
The vehicle-driver characteristics must be as rich as possible in
order to obtain the best risky group specification.

3.2. Method

The cross tabulation has been carried out from the common set of
variables of two files : an 1/10th accident file on national main
roads in 1974 and a driver file obtained from a sample survey of

service-stations situated on main road. The estimated bias to
vehicle autonomy is very weak. So we can consider that the number
of road-users stops at service-station are representative of their
driving distance.

We select for the driver four categorical variables : 12 socio-
professional categories, age with 7 categories, sex with 2 categories, driving licence age   with 5 categories. For the vehicle,
we select two categorical variables : make-type with 21 categories,
and model year with 5 categories. We add two more categorical data :
season with 3 categories and day/night with 2 categories.

The method used to analyse the cross-tabulated data is a segmentation algorithm. The aim is to obtain a hierarchical tree of
successive population partition in two sub-populations which are
the most heterogenous according to their risk. At each step the
partition is obtained by the dichotomy of variables categories
which gives the greastest chi square distance between the distributions of the population involved in an accident and the driver
population.

For a p categorical variable, the distribution of people involved
in an accident is

$$a_1, \ldots, a_p \quad \text{with } \Sigma a_p = 1$$

the driver people distribution is

$$c_1, \ldots, c_p \quad \text{with } \Sigma c_p = 1$$

We look for the dichotomy (I, II) of the categories i of this
variable which provides the maximal chi square distance. Let be
(a_I, a_{II}) and (c_I, c_{II}) the new distributions. Two distances can
be used :

a) chi square distance with reference to the driver population

$$d_1 = ||a_I - c_I||^2 c_I$$

$$= \frac{(a_I - c_I)^2}{c_I} + \frac{(a_{II} - c_{II})^2}{c_{II}}$$

$$= \frac{(a_I - c_I)^2}{c_I (1 - c_I)}$$

$$= (1 - r_I)^2 \frac{c_I}{1 - c_I} \quad \text{with } r_I = \frac{a_I}{c_I} \text{ as group I relative risk.}$$

b) Chi square distance with reference to the accident involved population

$$d_2 = ||a_I - c_I||^2_{a_I}$$

$$= \frac{(a_I - c_I)}{a_I (1 - a_I)}$$

$$(1 - \frac{1}{r_I})^2 \frac{a_I}{1 - a_I}$$

We did not compute all the possible dichotomies of the variable categories by imposing a constraint : the set of acceptable dichotomies is the one that respects the risk order. The variable categories are ranked according to their growing risk $\{r_1, ..., r_p\}$. The optimal cut according to the chi square distance is to be found between $(p - 1)$ instead of $\frac{p(p-1)}{2}$ combinations. This constraint is natural because we refuse the neutralisation of two extreme groups in a medium one. We did not prove that the optimal dichotomy giving the maximum chi square distance belongs to the ranked dichotomies set for more than three categories.

This algorithm has been computed with a stopping rule based on a 5 % threshold of accident involved or driver population in a class.

3.3. Results

According to the reference population, we obtain two trees (fig. and 2) which order variables and define dichotomised groups on their risk value. The chi square distance tends to make appear high risk marginal groups with driving population in reference and low risk groups with accident involved population in reference.

The most striking and newest output is the occurrence of socio-professionnal categories as the most predictive variable on which are linked driving licence age , age, make-type and season.

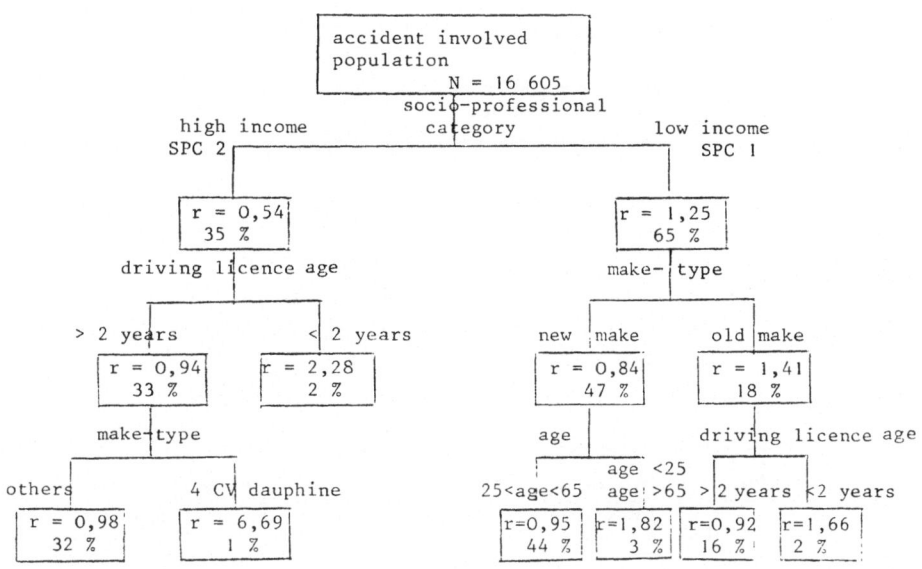

Fig. 1 Segmentation on accident involved population with r as
relative risk from the preceeding group and % as group frequency.

SPC 1 : farmers, craftmen, storekeepers, employees, workers, stu-
        dents, pensioners.

SPC 2 : manufacturers, merchants, high and middle-grade employees,
        learned profession, technicians.

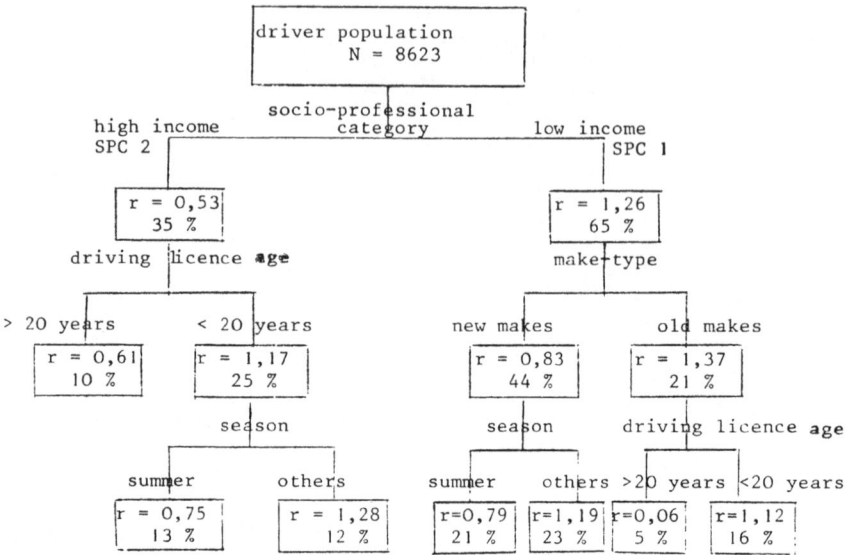

Fig.2. Segmentation on driver population with accident involved
       population in reference.

3.3. Conclusion

The segmentation is a method well fitted to relative risk study
without modelisation attempt. It is only a descriptive method, but
it provides outpouts easy to read : binary partition of population
and hierarchy of variables. But this simplification (partition in
two monothetic classes) comes with an information reduction. Moreover the resolution algorithm can be heavily time-spending with
great samples even   with the risk order constraint. A risk research
from an Insurance Companies file is actually in progress at ONSER.

4. EPIDEMIOLOGICAL STUDY

4.1. Introduction

This study [14] belongs to a primary safety sample survey where we
want from a detailed and real-time investigation of accident to
make a clinical evaluation of countermeasures. The proposed survey
requires a stratification of French Departments. It can be obtained by a typology based on some variables in harmony with the countermeasures.

4.2. Method

The baseline data are accidents of 1975 in all French departments but Paris and its suburb.

The categorical variables are the following :

- road : road and traffic category/road width/alignment/
- driver : age/driving licence age
- environment : intersection/urban or rural area/state of surface/ day-night/
- vehicle : category/age/
- accident : gravity/involved vehicle/

The categories of each variable are between 2 and 8. The I x J contingency table dimensions are 91 x 64 : I = {91 departments}, J = {64 variables categories} with as size k (i,j) the count of accident or the accident frequency for one department and one variable category.

We analyse this contingency table by two commonly used data analysis methods : the correspondence analysis joined with the hierarchical ascending clustering.

The correspondence analysis [1] with a chi square distance calculated on rows and columns provides eigen values, factors and factor graphs of the subjects or categories scatter diagram according to its inertial decomposition. The aim is to obtain a simplified representation of the scatter diagram in view to identify its pattern and discover relationships. What we try to hold and interprate is the divergence, gap with regard to independence hypothesis. This can be seen on the data reconstruction formula :

$$k(i,j) = k(i)\ k(j)\ \left[1 + \Sigma\ \{\lambda_\alpha^{-1/2}\ F_\alpha(i)\ G_\alpha(j)\ /\alpha\ \epsilon A\} \right]$$

where the size k(i,j) is equal to the margins product (with respect to independence hypothesis) plus a sum of terms composed of eigen values and products of factors on I and J.

The hierarchical ascending clustering [2] (C.A.H.) to be consistent with the correspondence analysis must work with the chi square distance calculated between subjects and with inter-class variance maximisation as agregation criterion.

This leads to a mutual enrichment of results from analysis and clustering.

4.3. Outputs   from correspondence analysis

The correspondence analysis is applied to the frequency contengency table. Each department has the same weight equal to 1/91.
If, instead of frequency, we use the number of accident, each
department will have a weight proportional to its accident score.
But the analysis outputs would be pratically the same because of
the great stability of the correspondence analysis.

The percentages of inertia explained by the first three axis are
important : 20 + 13 + 11 = 54 %. The identification of factors is
carried out with two indicators aid called absolute and relative
axis contributions.

The first axis is the substructure axis (fig. 3) constructed by
variables as road width, traffic category. On this axis are scaled
from the left to the right the different road width   classes from
- 5,75 m to + 11,5 m and also the number of lanes from two narrow
lanes to four separate lanes and motorways.

The second axis is the urban density axis (fig. 3) which oppose
small urban area to rural area. Associated to rural area, we find
main roads (routes nationales), when to urban area we find pedes-
trians and two-wheeled vehicles accidents.

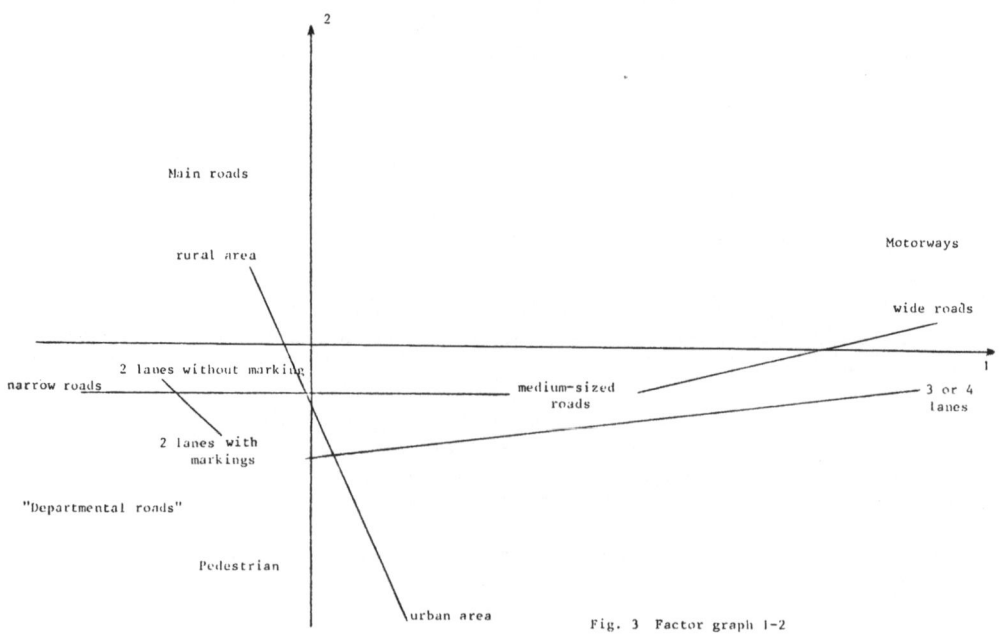

Fig. 3 Factor graph 1-2

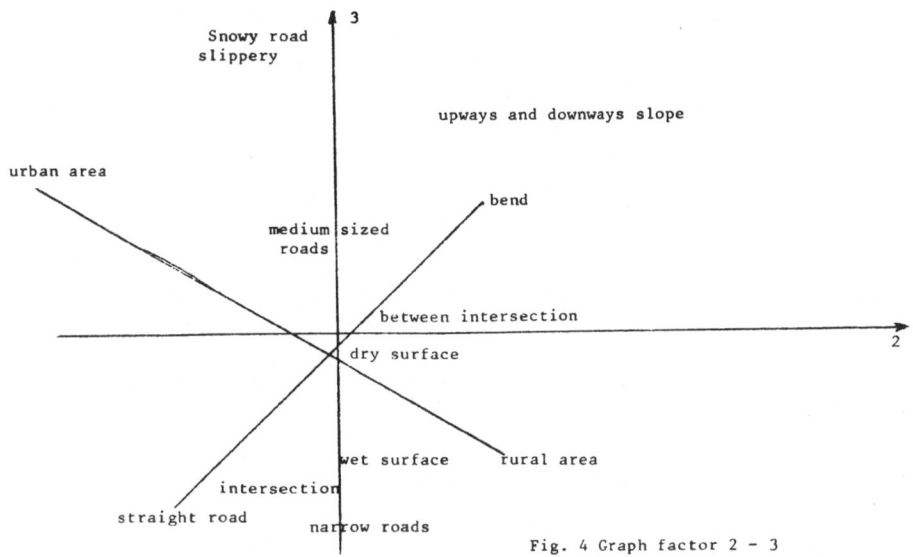

Fig. 4 Graph factor 2 - 3

The third axis is the <u>road physical characteristics axis</u> (fig. 4) with several opposition systems working together :

in intersection   between intersection
on straight road   on bend
on flat road   upways or downways slope

We find also weather items like snowy and slippery roads. Adverse weather conditions increase risk specially on mountain roads. Near this axis are the narrow rural roads which involve a lot of inter-sections.

A linear regression between accident factors and substructure, socio-economical and geographical variables would give good results because the first factor is highy correlated to substructure, the second one to socio-economical variables, and the third one to geographical (climate, region) variables. The analysis allows to isolate three uncorrelated factors which give a good idea of scatter diagram structure and of the correspondence between departments and accident characteristics : substructure, urban density, physical road characteristics. The others variables are of secondary impor-tance.

4.4. Departments classification

The C.A.H. algorithm applied on the same contingency table provides
the following tree (fig. 5) :

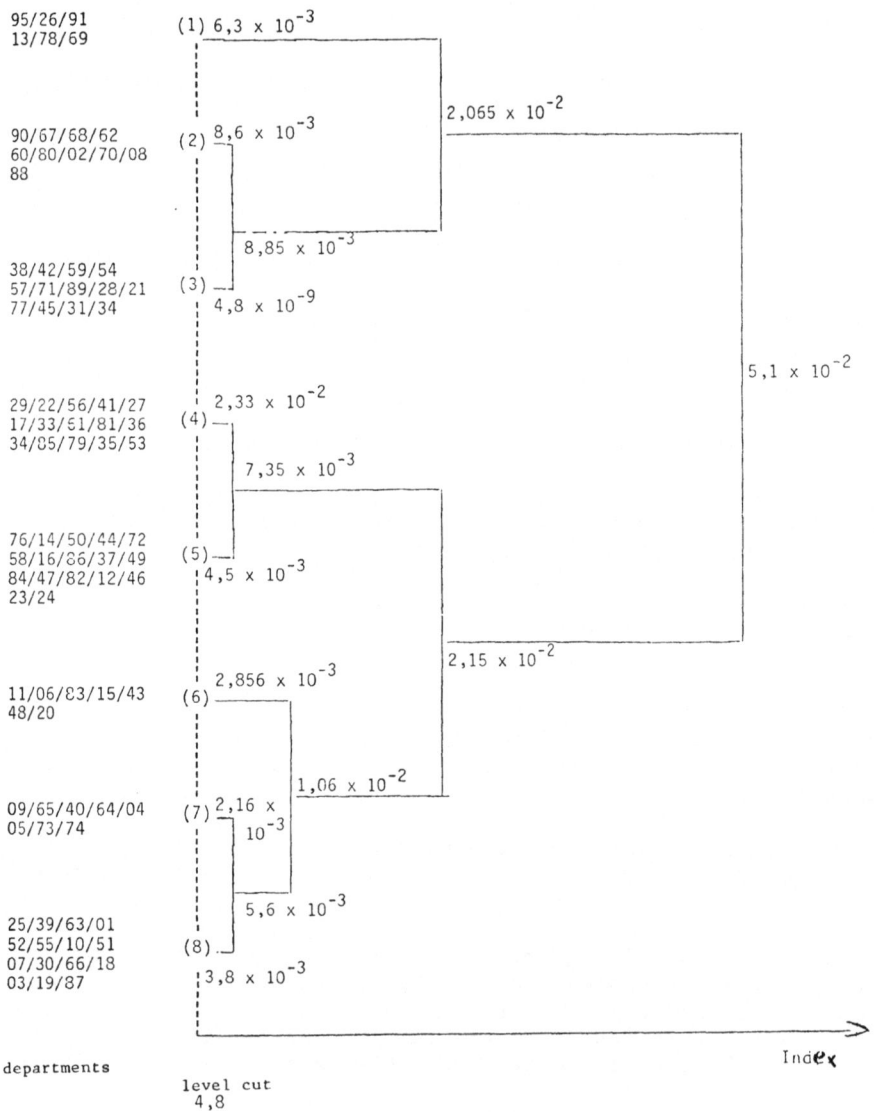

Fig. 5 Tree of departments

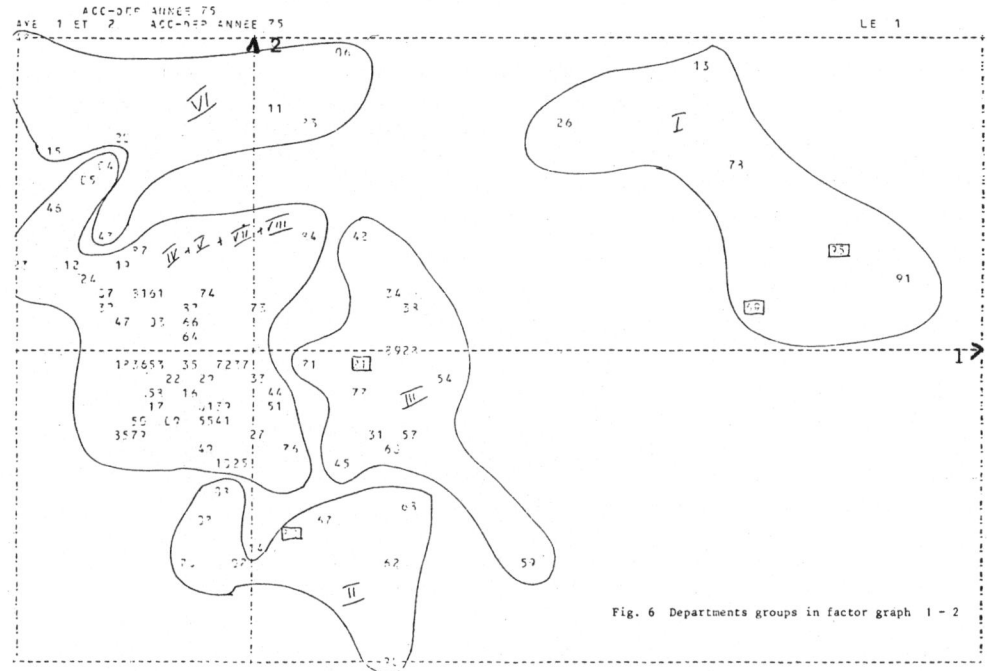

Fig. 6 Departments groups in factor graph 1 - 2

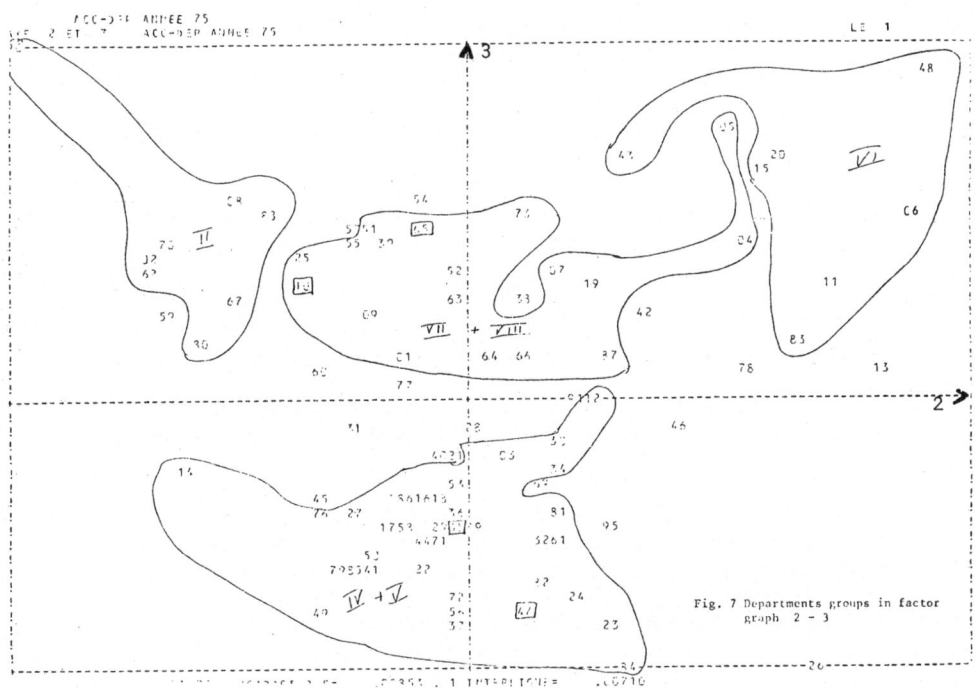

Fig. 7 Departments groups in factor graph 2 - 3

We shall retain eight groups which can be identified on factor graphs.

Groups 1, 2, 3, 6 clearly appear in graph 1-2 (fig. 6).
The third axis (fig. 7) lets to distinguish groups 4 and 5 opposed to groups (7, 8) and 6. The visualisation is not perfect because the clustering is done from the contingency table and not from the first three factors.

The identification of each group can be done from their position in factor graphs :

- group 1 : 6 departments near or with great urban area   like Paris, Lyon, Marseille ; crossed through by important traffic axis.

- group 2 : 10 industrial departments from North and East of France with an high density of small towns.

- group 3 : 13 departments with a medium density of small towns, crossed by high traffic main roads which served important regional centers like Dijon, Grenoble, Lille, Nancy.

- group 4 : 15 agricultural departments with a majority of rural roads situated in West and Center of France.

- group 5 : 17 agricultural departments with rural lanes and crossed by high traffic roads ; chained from Normandy to South West of France.

- group 6 : 7 departments with a lot of sinuous roads of South of France.

- group 7 : 7 departments from Alpes and Pyrénées gathered by particular geometry of roads and adverse winter conditions.

- group 8 : 15 departments with tortuous roads sensitive to adverse weather conditions situated in middle mountain region like Jura and Auvergne.

A cut at an apper level provides too a partition in six groups : 1, 2, 3, 4 + 5, 6, 7 + 8 which can be used as a basis for stratification.

The seasonal variations can be studied by the correspondence analysis of the (quarter x department) x variables contingency table where the temporal dimension is introduced. The first axis remains the substructure axis. The second axis is the seasonality axis which discriminate the quarter x department according to the

opposition winter + autumn / spring + summer. The seasonal effect
acts by means of two factors : day-light length and weather conditions which built this second axis. The third axis is a compo-
sition of urban density and road design axis. These four factors
have been taken into account to elaborate a sample survey design
which has not been realized for budgetary reasons.

4.5. Conclusion

Thanks to their mutual enrichment, the correspondence analysis
joined to hierarchical ascending clustering seems a good data
analysis tool specially with categorical or discrete variables
which are current in road safety field. We can also use them to
analyse large psycho-sociological surveys because of their synthe-
sis power, or with a regression goal to explore and detect linear
or non linear relationship between groups of variables, or more
with a pattern recognitive goal to make a typological research.

5. CONCLUSION

By exposing these three case-studies, we have tried to show that
the statistical treatment of contingency table can be very enri-
ching for road safety researches thanks to technics coming from
inferential statistics or data analysis. We do not need any special
methods but accurate and adaptable statistical ones which can be
exploratory as confirmatory methods and allows us to look at the
data, build and test a model, draw conclusions and begin again if
necessary.

BIBLIOGRAPHY

[1]   BENZECRI J.P. et son équipe. L'analyse des données
      Tome 2 : L'analyse des correspondances - DUNOD 1975

[2]   BENZECRI J.P. et son équipe. L'analyse des données
      Tome 1 : La taxinomie - DUNOD 1975

[3]   BIECHELER M.B. Cahier des méthodes statistiques applicables
      à la sécurité routière - ONSER 1973

[4]   CAMBOIS M.A. et FONTAINE H. Hiérarchisation de certains
      facteurs accidentogènes suivant l'importance de leur incidence
      sur le risque - ONSER 1979

[5]   CAMPBELL D.T., STANLEY J.C. Experimental and quasi-experi-
      mental designs for research
      Rand Mac Nelly Publishing Co, Chicago, 1963

[6]   CARROLL P.S. Classifications of driving exposure and accident
      rates for highway safety analysis
      Accident Analysis and Prevention Vol 5, n° 2, 1973

[7]   FLEURY D. Les glissières implantées sur terre-plein central
      ONSER 1974

[8]   FLEURY D. Les glissières implantées devant les plantations
      d'alignement - ONSER 1974

[9]   GARDWOOD and NEWBY Utilisation du χ^2 pour la comparaison
      des fréquences d'accidents. Symposium OCDE PARIS 1970

[10] LASSARRE S. Efficacité des nouvelles normes de marquage
      ONSER 1976

[11] LASSARRE S. Etude expérimentale de l'efficacité des délinéa-
      teurs - ONSER 1976

[12] LASSARRE S. Etude expérimentale de l'effet du rainurage du
      béton sur autoroute - ONSER 1975

[13] LASSARRE S. A propos des tests statistiques sur variables poissoniennes, utilisés dans le domaine de la sécurité routière Revue de Stat. Appl. 1977 vol XXV n° 3

[14] LASSARRE S. Typologie des départements selon les caractéristiques des accidents - ONSER 1977

[15] TARRANTS W., VEIGEL C.H. The evaluation of highway traffic safety programs. A manual for Managers - NHTSA 1977

MINIMUM DISCRIMINATION INFORMATION WITH LOG-LINEAR MODELS: A MORE ROBUST ANALYSIS

H.E. Roland

University of Southern California, Los Angeles, California

ABSTRACT

This paper compares the analysis of frequency data of bicycle/motor vehicle accidents when analyzed by the minimum discrimination approach with log-linear models with the more conventional approach using a two-way contingency table and the Pearson chi-square. It shows that the minimum discrimination approach provides a more robust analysis of the factor relationships.

I. INTRODUCTION

This short paper is concerned with illustrating the advantage of the principle of minimum discrimination information (MDI) for statistical model building with log-linear models over that of the more traditional nonparametric form of data analysis. The mathematical development will be deferred to papers by Kullback (1974) or Gokhale and Kullback (1978).

The traditional methods of analyzing frequency data are one of three approaches: (1) regression, (2) analysis of variance, (3) Pearson chi-square, the first two being parametric and the latter non-parametric. The Pearson chi-square approach is useful in that it requires few assumptions about the underlying data population, however it provides a less than robust separation of the variable effects. The first two

methods require assumptions which are difficult to
establish.

(1) Random sampling
(2) A specified structure of the population means
(3) Constant variance
(4) Normality of the distributions of the factors

The MDI approach illustrated herein requires only
random sampling, the most fundamental of statistical
assumptions which must be granted when sampling from
major real-world systems.

This paper compares the results of the MDI approach
with that of the Pearson chi-square contingency table
approach.

Discussion

The type of problem being addressed by the MDI approach
is that of independent observations classfied into cells
according to categories and levels of categories. Such
data are typical of traffic safety data. In this class
of problems, data are gathered concerning a particular
safety situation. It is acquired in accordance with
some previously determined experimental design in which
the factors or independent variables have been selected
to illustrate their effect on the dependent variable.
The dependent variable may be a direct measure of frequency or severity of the accident or it may be an indirect
measure which will, if allowed to persist without corrective
action, result in an accident.

This paper will consider two data sets from a
Department of Transportation research project (Roland
1979) which are used to examine the parameters of motor
vehicle/bicycle accidents. Data were gathered by research accident investigators using an in-depth investigation with emphasis on the human factors.

The original research effort utilized the Pearson
chi-square statistic to analyze contingency tables of
the various data sets. The results of this analysis will
be compared to the analysis of the same data sets using
the MDI approach.

Table 1 below shows a data set which examines

TABLE 1

BICYCLIST SEARCH/AGE GROUPS

Bicyclist function failures

Bicyclist age	Ab search No.	%	Bd detection No.	%	Cb evaluation No.	%	Db decision action No.	%	Eb human action No.	%	Eb bicycle action No.	%	Row Total No.	%
0-10	75	38.5	32	28.8	12	7.0	2	20.0	3	12.5	4	10.5	128	23.3
11-15	66	33.8	41	36.9	76	44.2	1	10.0	15	62.5	19	50.0	218	39.7
16-25	36	18.5	25	22.5	51	29.7	4	40.0	3	12.5	13	34.2	132	24.0
26-35	8	4.1	9	8.1	17	9.9	1	10.0	2	8.3	1	2.6	38	6.9
36-50	2	1.0	2	1.8	6	3.5	0	0	0	0	0	0	10	1.8
51+	8	4.1	2	1.8	10	5.8	2	20.0	1	4.2	1	2.6	23	4.2
Column total	195	35.5	111	20.2	172	31.2	10	1.8	24	4.4	38	6.9	549	100.0

$X^2 = 77.33862$ with 25 d.f. $p = 0.0000$

bicyclist function failures grouped with a bicycle
action failure versus bicyclist age.

It can be seen that there is a statistically
significant effect of age on function failure.
However, because of the six by six size of the contingency table, the analysis is less than robust.   It
is not clear without further analysis where the
majority of the age effect lies or what is the age
which provides the predominate effect.

The MDI approach with the log-linear model is
developed in the previous two references and by
Gokhale (1979).   Briefly the method is to define a
function:

$$I\ (p:\pi) = \overset{\omega}{\Sigma} p(\omega) \ln \frac{p\omega}{\pi\omega} \qquad\qquad (1)$$

where: $p(\omega)$ and $\pi(\omega)$ are any two probability
distributions defined on any set of elements
$(1,2,3,\ldots\Omega)$.

The solution is then to select from all
probability distributions the one p* which satisfies
equation (2) below and is closest to $\pi(\omega)$.

$$Cp = \theta \qquad\qquad (2)$$

where :   C is a $(r + 1) \times \Omega$ matrix and θ is a
$(r + 1) \times 1$ vector.   Both C and θ are assumed known.
The first row of C consists of all ones and the first
element of θ is also one.

The solution is found by minimizing the function
$I(p:\pi)$.   The constraint is provided by the hypothesis
of the particular analysis in question.

The cumbersome computation of the method is
done by a computer program which is available with
documentation.   This program is known as CONTAB.

Table 2 below is the summary table of the MDI
approach to the data set of Table 1.   The form of
the solution is the external constraint approach.

TABLE 2

MDI HYPOTHESIS

Characteristic	Index	Value
Bicyclist Age	h	0-10, 11-15, 16-25 26-35, 36-50, 51+
Bicyclist Function Failure	i	Search, Detection, Evaluation, Decision, Human, Bicycle Action

Hypothesis Testing

	I.S.	df	Prob.
(a) x (h .), x (. j)	183.6	25	.0000

It is apparent that factor h is not independent of factor i. What is of interest is which level of factor h is the predominant factor in the relationship.

The Pearson chi-square analysis had found that the relationship between the 0-10 age group and the search failure was the major relation. However, this proves to be an outlier in the MDI approach and the major effect is the relationship between the 11-15 age group and the search failure. Thus, when the model of the MDI approach fits, the previously identified age group level is cast out in favor of quite another age group.

Table 3 below is a table describing the relationship between bicycle speed, bicycle type, and accident severity. A three-way table is not well suited to a Pearson chi-square analysis. From this table it is difficult to determine which is the predominate influence on severity, bicycle speed or bicycle type.

TABLE 3

INJURY SEVERITY/SPEED/BICYCLE TYPE

Speed	Severity	Bicycle type					
		Lightweight		Medium-weight		Conventional or junior	
		No.	%	No.	%	No.	%
0-15 mph	None, moderate	39	97.5	228	88.4	115	89.1
	Severe, fatal	1	2.5	30	11.6	14	10.9
16-25 mph	None, moderate	5	71.4	29	72.5	5	23.8
	Severe, fatal	2	28.6	11	27.5	16	76.2
26 mph & over	None, moderate	2	66.7	16	64.0	5	31.3
	Severe, fatal	1	33.3	9	36.0	11	68.8

$x^2 = 23.01$ with 6 d.f.   p < .001

Table 4 below is the MDI approach with the external constraint solution.

TABLE 4

MDI HYPOTHESIS

Characteristic	Index	Value
Speed	h	0-15, 16-25, 26 + ,
Bike type	i	Light, Medium Conventional-Junior
Severity	j	None-Moderate, Severe Fatal

Hypothesis Testing

	I.S.	df	Prob.
(a) x (h,i .),x(..j)	74.3	8	.0000
(b) x (h,i .),x(h . j)	73.2	6	.0000
(a:b); h i ; j	1.1	2	.4231
(c) x (h,i .),x(. i,j)	12.8	6	.0478
(a:c), i,j: j	61.5	2	.0000

Examining Table 4 above, hypothesis (a) tests for the independence of severity, speed, and bike type. This hypothesis is clearly rejected.

Hypothesis (b) examines the interaction between speed-bike type and speed-severity. It is not surprising to find a dependency.

When hypothesis (b) is nested in hypothesis (a) for the effect of speed it is quite surprising to find little effect of speed.

Hypothesis (c) examines the interaction between speed-bike type and bike-type and severity. As would be expected there is an interaction. However, when hypothesis (c) is nested in (a) to develop the effect of bike type it is found that the bike type effect is significant.

This result is somewhat different than that which would be deduced from Table 3 which indicates only a statistical relationship between the variables but the exact nature of the relationship is difficult to determine.

In addition to establishing bike type as the dominant influence on injury severity when compared to speed, the MDI program shows which of the data points were outliers or so far outside the fitting process as to be incapable of being fitted. Thus,we see that the speed classes of 0-15,in conjunction with light and conventional-junior bicycle types, are discarded as outliers when selecting the bicycle type as the most important variable in determining injury severity.

An examination of the log-ratio, an output of the CONTAB program, allows us to determine that the medium bicycle is the bicycle type which is statistically significant in the influence on injury severity.

CONCLUSIONS

We may conclude that the MDI approach with the log-linear model provides a challenge to the traditional nonparametric analysis. It seems to provide a more robust analysis of safety data. It is of particular importance when analyzing n x m tables with n and m greater than three. The MDI approach also provides a more robust analysis when more than two factors are being considered. The method has the additional advantage of identifying relationships between the factors which may simply be due to bad data.

REFERENCES

1. Gokhale, D.V.; Kullback, S. Information and
    Contingency Table. Marcel Dekker, Inc., N.Y. 1978

2. Kullback, S.; and Ku, Harry H. "Loglinear Models
    in Contingency Table Analysis." The American
    Statistician 28, No. 4, November 1974, pp 115-122

3. Gokhale, D.V. "Analysis of Ecological Frequency
    Data: Certain Case Studies." Technical Report No. 51,
    University of California at Riverside, Department of
    Statistics, February 1979.

4. Roland, H.E. "Investigation of Motor Vehicle/
    Bicycle Collision Parameters." Vols. I and II,
    Traffic Safety Center, University of Southern Cali-
    fornia, NHTSA, 1979

PRELIMINARY ANALYSIS OF THE NATIONAL CRASH SEVERITY STUDY: FACTORS IN FATAL ACCIDENTS†

Susan Partyka

National Highway Traffic Safety Administration,
U.S. Department of Transportation

I. INTRODUCTION

Purpose

This study examines the fatalities on the National Crash Severity Study (NCSS) and compares them to (1) fatalities on the Fatal Accident Reporting System (FARS); and (2) those occupants on NCSS who were not fatally injured. Many factors can influence the likelihood of occupant injury and death in an accident. In particular, the speed of the vehicle at impact and the seating location and age of the occupants are expected to affect the chance of surviving the accident.

Several other characteristics also were found to be associated with the incidence of fatality in traffic accidents. These include the number of vehicles involved in the accident, the direction of the impact force, the change in vehicular speed during impact, and such occupant characteristics as seating position, age, sex, ejection status, and restraint use.

The findings of this report are not intended to be a description of the national experience, but only of the occupants of tow-away accidents in the areas sampled by NCSS.

---

† This paper was originally prepared for the Mathematical Analysis Division, National Center for Statistics and Analysis, National Highway Traffic Safety Administration, U.S. Department of Transportation: DOT HS-804 773, June 1979. The original may be obtained from the National Technical Information Service, Springfield, Virginia 22161.

The National Crash Severity Study

The National Crash Severity Study (NCSS) is sponsored and conducted by the National Center for Statistics and Analysis (NCSA)
of the National Highway Traffic Safety Administration (NHTSA).
Seven NCSS teams investigate traffic accidents in eight sampling
areas. These areas were chosen to approximate the national characteristics from areas which had available, experienced accident
investigators. These sampling areas were a judgment sample — they
were not chosen randomly and may not be representative of the accidents occurring nationwide. However, the sites are scattered
throughout the country and include both rural and urban areas.
Within each area police-reported tow-away accidents are investigated following a rigorous sampling plan. The tow-away criterion includes only vehicles towed because of damage — those towed
not for the convenience of the driver or because required by
State law. Within this frame, all accidents with a fatality or
an injury with the resultant overnight hospitalization are investigated. Less severe accidents are investigated at one out of
four if an occupant was transported to a medical facility, and
one out of ten if no transport was involved in the accident. This
sampling plan yields a higher proportion of severe accidents than
normally would be found in these areas. To adjust the figures to
the correct proportions, as would occur if every accident had
been investigated, each accident is multiplied by the inverse of
the sampling fraction. Since only one-tenth of all accidents
with no transport are investigated, each of these is multiplied
by ten, reflecting a higher incidence of low-injury accidents.
This process results in a "weighted" file which, while attempting
to estimate correctly the frequency of accident factors in the
overall NCSS accident population, also may be skewed by one or
two unusual cases when the number of cases is small. This does
not affect the fatal accidents on the file, which are sampled at
100 percent.
Sources of information on NCSS are "The National Crash Severity
Study" (C. Kahane, R. Smith and K. Tharp) and J. Hedlund's "A Working Guide to the National Crash Severity Study". (References 1
and 2).

The Fatal Accident Reporting System

The Fatal Accident Reporting System (FARS) contains data on
the census of fatal accidents occurring within the fifty States
and the District of Columbia. This provides the only nationwide
file of data on traffic accidents. FARS analysts use official
State records to gather information on fatal traffic accidents.
Traffic fatalities resulting from non-accident causes, such as
suicide with a motor vehicle, are excluded from the file.
The FARS data collection forms differ from the NCSS forms.
While FARS is limited to information available on State records,

NCSS uses medical records, interviews, and on-site accident investigators to gather additional information. Therefore, many of the
data in NCSS, such as specific injuries and measurements of vehicle damage, are not available on FARS.

In particular, FARS does not distinguish reasons for towing the
vehicle, and in fact some States require the towing of all vehicles
involved in a fatality.

The FARS Annual Report, produced by the NCSA, is a good source
of information on this nationwide census of fatal traffic accidents. In addition, special studies on topics of current safety
interest are published from time to time (Reference 3).

Analyses

As of December 1978, the NCSS file was approximately 93 percent complete for the months of January 1977 through March 1978.
Thie file contained data on 6,216 accidents (29,919 when weighted),
of which 372 had a fatality, and 13,525 occupants of towed passenger cars (58,069 weighted) of whom 442 were fatalities. The 1977
FARS file was almost 100 percent complete and contained 21,880
fatal accidents involving tow-away passenger cars, and 25,818 occupant fatalities in tow-away cars. The first quarter (January
through March) of the 1978 FARS data was approximately 100 percent complete. Thus, each of the two accident files (NCSS and
FARS) was fairly complete for the fifteen-month period, January
1977 through March 1978.

The analyses of this study involved computing distribution of
accident, vehicle, and occupant characteristics for NCSS fatal
accidents and comparing these to (1) similar distributions for
FARS fatal accidents, and (2) distributions for all NCSS accidents.
The Chi-square goodness-of-fit test was used to help evaluate the
differences between NCSS and FARS fatality distributions. This
test was an aid in assessing if these differences could be considered real ("significantly large"), or were small enough to be
attributable to chance.

The comparison of NCSS fatalities with all NCSS occupants required estimating conditional probabilities of fatality. This
was estimated as the rate of fatality under various conditions,
such as the percentage of people involved in rural accidents who
are killed. In addition, the iterative model-fitting program
CONTAB was used as a tool to separate the effects of correlated
pairs of fatality factors, such as rollover and ejection.

The results of this study pertain only to the NCSS sampling
areas. To extend these results beyond NCSS it would be necessary
to consider, among other things, (1) whether NCSS accidents are
representative of accidents in other areas, (2) the effect of the
seven percent of NCSS data that is not yet on the fifteen-month
file, and which may differ from the 93 percent which has been
automated, and (3) the necessity of limiting the NCSS study to
tow-away accidents. For these reasons, the reader should exercise

caution in using NCSS distributions to estimate the incidence of accident factors in all national fatal traffic accidents.

Summary of Results

The distributions for several important variables appear to be similar on the FARS and NCSS files. The two files have approximately the same proportions of rural vs urban accidents, large vs small cars, and front vs side damage to the vehicle in which a fatality occurred. Fatally injured occupants on the FARS and NCSS files have similar distributions for seating area, location, sex, and restraint use.

There are, however, some significant differences between the files. FARS has more single-vehicle accidents than does NCSS (48% vs 41%), more rollovers 26% vs 20%), older fatalities (43% vs 38% are 30 years or older), and fewer ejections (24% vs 29%). Section II of this report studies these variables in more detail and suggests explanations that involve the differing sampling areas and coding practices of the investigators.

When comparing these two data files, the differing definitions, codes, investigative methods, and sources of accident information must be considered. The source of FARS data is police reports and other official documents which are already prepared and available. On the other hand, NCSS uses in-depth investigations which include physical evidence and the accident scene. These investigations are specifically designed to gather certain information considered essential for accident analysis, but not generally available through record searches.

Within the NCSS sampling areas, some factors are associated with a much higher fatality rate for the occupants involved, as discussed in detail in Part 3. The following are some of the important contrasts which were found in the analysis. The accident factors, the contrasting fatality rates for each pair of factors, and the incremental difference between the pair of factors are shown here. The actual counts of the data, from which these rates are computed, are presented in Section II of this report.

Factor	Rates of Fatality	Rate Increment
Accident Level:		
Single vs multi-vehicle	0.01315 vs 0.00588	2.2
Rural vs urban	0.02007 vs 0.00397	5.1
Incidence of fire: yes vs no	0.13514 vs 0.00712	19.0
Fixed object vs car/vehicle	0.01261 vs 0.00633	2.0

Factor	Rates of Fatality	Rate Increment
Vehicle Level:		
Size: small vs large	0.00878 vs 0.00756	1.2
Rollover vs non-rollover	0.03209 vs 0.00756	4.2
Damage area: side vs front	0.01153 vs 0.00653	1.8
Damage area: front vs back	0.00653 vs 0.00074	8.8
Delta V: over 12 vs up to 12 mph	0.01725 vs 0.00049	35.2
Occupant Level:		
Location: front vs second seat	0.00769 vs 0.00580	1.3
Seat area: window vs middle	0.00766 vs 0.00529	1.4
Age: over 30 vs up to 30 years	0.00807 vs 0.00750	1.1
Male vs Female	0.00909 vs 0.00572	1.6
Restraint used: no vs yes	0.00545 vs 0.00313	1.7
Ejected vs not ejected	0.19856 vs 0.00480	41.4
Entrapped vs not entrapped	0.19554 vs 0.00469	41.7

The column "Rate Increment" shows, for example, that victims of rollovers have over 4 times the fatality rate of occupants of non-rollovers. The factors with the highest incremental difference are entrapment, ejection, delta V, and incidence of fire in the accident. Other factors, with a lower rate increment, still substantially increase the estimated probability of fatality. Most important, from the point of view of the occupant, is that unrestrained occupants suffered 1.7 times the fatality rate of restrained occupants. As contrasted with age or damage area of the vehicle, each occupant has ultimate control over the choice of this factor.

Some factors are interrelated to the degree that contingency table analysis is a useful way to sort out the effects. This is explained in Section IV. The most important result is the quantification of each of the factors "rollover" and "ejection". Since these two occurrences are correlated, special techniques are needed to estimate how much effect each factor has separately on the fatality rate. The conclusions of the analysis are that a roll-over, for each ejection status, increases the fatality rate by a factor of 2.3. Similarly, an ejected person has an increased risk of almost 40 times that of an unejected person, for each of the categories rollover and non-rollover.

These estimates of the relative risk to the occupant will be used to help define the major causes of accidents, and suggest important areas for further research. When an assessment of NCSS representativeness is complete, this type of analysis will be used not only to define the conditions with the greatest risk to the individual occupant, but also the most frequent causes of injury and fatality. This combination of knowledge will help to determine the priorities of highway safety research.

II. COMPARISON OF NCSS AND FARS FATAL ACCIDENTS

Approach

Preliminary studies of the NCSS data have shown eleven variables to be of major importance in predicting injury severity. (References 4 and 5). In evaluating the NCSS fatality file, the distributions of these elements were compared to the corresponding distributions on the FARS file. The chi-square criterion at the 5 percent level was used to help separate small differences between the files from significantly large ones. Further studies will attempt to make national estimates based upon the NCSS file, and adjustments will need to be made when the file is not consistent with the FARS data.

The distributions of the number of vehicles invovled; vehicle rollover and damage description; and occupant age and ejection status were significantly different. However, the rural/urban classification of the scene, size of the vehicle, occupant seat area and location, and the occupant sex and restraint use each had distributions which were consistent with those on the FARS file. These differences were small enough to be the result of chance.

The following sections compare the two files in more detail for each of these accident factors.

Accident Description

Table 1 shows the higher proportion of FARS accidents which involve only a single vehicle: 48 percent, as contrasted with only 41 percent of the NCSS fatal accidents. This is a significant difference, using the chi-square criteria at the 5 percent level. The probability (P) of a larger difference for normally-distributed, independent variables is only 0.008. Thus, the proportion of single-vehicle accidents appears to be very dependent

TABLE 1

Accident Description Variables:
Counts and Percentages of Known Data
for Fatalities only

NCSS Data

Number of Vehicles	Count	Percentage
One	154	41.4
Two	193	51.9
Three or more	25	6.7

Rural/Urban	Count	Percentage
Rural	223	59.9
Urban	149	40.1

Total accidents: 372
Known vehicles: 372 (100.0% of total)
Known rural/urban: 372 (100.0% of total)

FARS Data

Number of Vehicles	Count	Percentage
One	12,637	48.3
Two	11,972	45.8
Three or more	1,560	6.0

Rural/Urban	Count	Percentage
Rural	16,147	61.9
Urban	9,927	38.1

Total accidents: 26,172
Known vehicles: 26,168 (100.0% of total)
Known rural/urban: 26,074 (99.6% of total)

upon which file is used as the source of the information.
   One possible explanation for this difference could involve differences in the urbanization of the sampling areas of the two studies, because rural accidents are more frequently single-vehicle accidents than are urban accidents. The FARS file has a slightly higher proportion of rural accidents (61.9%) than does the NCSS accident file (59.9%). This difference was not found significant using the chi-square decision rule ($P = 0.435$) and is not large enough to account for the higher frequency of single-vehicle accidents. In fact, the FARS rural accidents involve a single vehicle 50% of the time, as contrasted with a rate of only 45% for

for NCSS. Similarly for urban accidents, 45% on FARS are single-
vehicle, as opposed to a rate of only 36% for NCSS. Urbanization
does not appear to be the major cause of the difference between
the two files in the number of vehicles involved. The FARS rate
of single-vehicle involvement is higher than that of NCSS, regardless of urbanization.

Vehicle Size and Damage

    The size of the vehicle in which at least one fatality oc-
curred is shown for NCSS and FARS in Table 2. The wheelbase is
used to define the six size categories of minicar (up to 94 inches
long) through large (over 123 inches long). Wheelbases for vehicles in 1978 FARS cases are not currently available, so 1977
figures are used. Almost 50 percent of the wheelbases on NCSS,
and 39% on FARS for 1977 cannot be uniquely determined from the
vehicle identification number (VIN), and are therefore unknown.
    The FARS file has a higher proportion of small cars through
the compact category than does NCSS (34.6%). However, the large
proportion of unknown makes it hard to determine if this difference is large enough to be considered real. The chi-square test
does not assess the difference to be significantly large ($P = 0.136$).
    While the NCSS file is based upon vehicles towed from the
scene because of damage, the closest thing in FARS is the variable "towaway", which includes vehicles towed for reasons other
than disabling damage. However, the NCSS file contains only one
case of a person who was killed in a non-towaway vehicle. This
indicates that most fatal accidents involve a great deal of damage
to the vehicle as well.
    Using the rollover variable present on the 1978 FARS forms,
25.8% of vehicles which contained a fatality also rolled over in
the course of the accident, as shown in Table 3. For NCSS, this
variable is not available, and rollovers are determined by a type
of damage distribution of "rollover" for either of the two most
severe impacts. The different ways of describing a rollover may
in part account for the significantly lower rate of fatal rollovers on NCSS ($P = 0.008$) — only 19.6% of the vehicles involved
a fatality. This lower rate is also consistent with the lower
proportion of single-vehicle accidents on NCSS, because 90 percent of rollovers do not involve another vehicle.
    For planar, non-rollover accidents, the FARS variable, "prin-
cipal impact point", is not directly comparable to the "area of
damage" recorded on the NCSS file. While FARS uses twelve clock
positions to locate the damage, NCSS uses the less precise desig-
nation of front/back/left/right to record which side was damaged.
The FARS variable was converted into the NCSS damage variable
and the results of these calculations are shown in Table 3. For
these non-rollover accidents, the ratio of front damage: side
damage is similar for the two files. For NCSS, the ratio is

TABLE 2

Vehicle Size as Determined by Wheelbase:
Counts and Percentages of Known Data
for Fatalities only.

NCSS Data

Size, Wheelbase	Count	Percentage	Cumulative Percentage
Minicar (up to 94")	6	3.1	3.1
Subcompact (95-102")	30	15.7	18.8
Compact (103-110")	30	15.7	34.6
Intermediate (111-117")	68	35.6	70.2
Full Size (118-123")	41	21.5	91.6
Large (over 123")	16	8.4	100.0

Total vehicles: 381
Known size: 191   (50.0% of total)

FARS Data

Size, Wheelbase	Count	Percentage	Cumulative Percentage
Minicar (up to 94")	1,006	7.3	7.3
Subcompact (95-102")	2,217	16.1	23.5
Compact (103-110")	2,254	16.4	39.9
Intermediate (111-117")	4,224	30.8	70.6
Full size (118-123")	2,852	20.8	91.4
Large (over 123")	1,183	8.6	100.0

Total vehicles: 22,425
Known vehicles: 13,736 (61.3% of total)

160:119 = 1.3; the FARS ratio is 6,026:4,943 = 1.2. The largest
differences between the damage distributions appear to be the result of the different coding practices for the rollover situation.

TABLE 3

Vehicle Damage Description:
Counts and Percentages of Known Data
for Fatalities only

NCSS Data

Vehicle Damage	Count	Percentage
Rollover	69	19.6
Non-rollover	283	80.4
Back	2	0.6
Front	160	45.5
Left	55	15.6
Right	64	18.2
Other known	2	0.6

Total Vehicles: 381
Known Damage: 352   (92.4% of total)

FARS Data

Vehicle Damage	Count	Percentage
Rollover	4,005	25.8
Non-rollover	11,531	74.2
Back	296	1.9
Front	6,026	38.8
Left	2,633	16.9
Right	2,310	14.9
Other known	266	1.7

Total Vehicles: 15,743
Known Damage: 15,536 (98.7% of total)

Occupant Factors

The seating positions of fatalities are shown in Table 4. Approximately two-thirds of these people were in the driver's seat at the time of the accident: 68.1 percent of NCSS fatalities were left front seat passengers (drivers), as were about the same number, 62.2 percent, in FARS ($P = 0.752$).

The rows and columns do not always add to the total shown because of the small number of people with a seat area which is other than those shown (e.g., "lying across seat") or unknown.

The distributions are not significantly different: approximately 92 percent of fatalities are in the front seat ($P = 0.454$) and

TABLE 4

Occupant Seating Position:
Counts and Percentages of Known Data
for Fatalities only

NCSS Data

Location	Left	Seat Area Middle	Right	Total
Front	292 (68.1%)	10 (2.3%)	91 (21.2%)	394 (91.4%)
Second	11 (2.6%)	6 (1.4%)	18 (4.2%)	36 (8.4%)
Third	- -	1 (0.2%)	- 1	1 (0.2%)
Total	303 (70.6%)	17 (4.0%)	109	(25.4%)

Total occupants   442
Known seating: 429 (97.1% of total)

FARS Data

Location	Left	Seat Area Middle	Right	Total
Front	19,825 (67.2%)	647 (2.2%)	6,789 (23.0%)	27,276 (92.4%)
Second	858 (2.9%)	402 (1.4%)	971 (3.3%)	2,236 (7.6%)
Third	5 (0.0%)	5 (0.0%)	2 (0.0%)	13 (0.0%)
Total	20.688 (70.1%)	1,054 (3.6%)	7,762 (26.3%)	

Total occupants: 30,855
Known seating: 29,504 (95.6% of total)

and 70-71 percent (for FARS and NCSS, respectively) are in a left side seat ($P = 0.823$).
    The joint distribution of age and sex is shown in Table 5. The proportion by sex is not significantly different between the

TABLE 5

Occupant Age and Sex:
Counts and Percentages of Known Data
for Fatalities only

NCSS Data

Age	Sex		Total
	Male	Female	
Under 20	87 (19.9%)	39 (8.9%)	126 (28.6%)
20-29	104 (23.8%)	42 (9.6%)	146 (33.1%)
30-39	46 (10.5%)	11 (2.5%)	57 (12.9%)
Over 39	66 (15.1%)	42 (9.6%)	108 (24.5%)
Total	305 (69.2%)	136 (30.8%)	

Total occupants: 442
Known age and sex: 437 (98.9% of total

FARS Data

Age	Sex		Total
	Male	Female	
Under 20	5,215 (17.0%)	2,680 (8.7%)	7,896 (25.7%)
20-29	7,044 (22.9%)	2,482 (8.1%)	9,526 (1.0%)
30-39	2,514 (8.2%)	1,137 (3.7%)	3,651 (11.9%)
Over 39	5,863 (19.1%)	3,780 (12.3%)	9,643 (31.4%)
Total	20.722 (67.2%)	10.131 (33.8%)	

Total occupants:
Known age and sex: 30,715 (99.5% of total)

two files (P = 0.374); approximately one-third of the fatalities
are female. However, the NCSS fatalities are significantly younger than their FARS counterparts (P = 0.021). While 62 percent of
the people killed within the NCSS study area are under 30, this
group accounts for only 57 percent of the FARS census of fatalities. A look at the under 30 group by sex shows that both males
and females are slightly younger in the NCSS study.

The NCSS data shows that restraint use, as determined by the
accident investigator, is about 5.4 percent for fatalities.
Table 6 contrasts this with the FARS usage of 4.1 percent. (If
only the States which report usage are considered, FARS also shows
that 5% of the fatalities were restrained). A very small proportion of those killed in passenger cars were using any form of restraint, and the difference between the two files is not significant at the chi-square 5 percent level (P = 0.230).

Using the chi-square test leads to the conclusion that there
is a difference in terms of ejection (P = 0.042). While 24.2 percent of the fatalities on FARS were partially or totally ejected,
28.7 percent of the NCSS fatalities whose ejection status was
known were ejected to some degree. The rate for both total and
partial ejection is higher in the NCSS sampling areas for the
known data.

However, there is a much higher rate of unknown for NCSS ejection status: for 13.3 percent of the fatalities it is not known
whether or not the victim was ejected. This would be sufficient
to account for the difference: if all of these people were actually not ejected, the two distributions, for NCSS and FARS, would
be similar. Computing ejection rates as percentages of total occupants (instead of using only occupants with known ejection stat-
tus as the base) results in estimates which are similar for the
FARS and NCSS fatalities. The rates for FARS and NCSS, respectively, are:

   (1) Total ejection: 20.2% vs 21.8%

   (2) Partial ejection: 3.5% vs 3.8%

   (3) No ejection: 74.3% vs 61.8%

   (4) Unknown ejection
       status: 1.9% vs 13.3%

A possible explanation of the difference in the rates of ejection
involves the different methods for determining that no ejection
took place. A tendency on the part of the NCSS analysts to require more proof that no ejection took place would account for
the different rates on the two files.

Comparison Summary

The NCSS fatal file has distributions which are similar to
FARS for many important variables. The description of the acci-

TABLE 6

Occupant Restraint Use and Ejection Status:
Counts and Percentages of Known Data
for Fatalities only

NCSS Data

Restraint Used	Count	Percentage
Yes	21	5.4
No	371	94.6

Ejection Status	Count	Percentage
None	273	71.3
Total	93	24.3
Partial	17	4.4

Total occupants: 442
Known restraints: 392 (88.7% of total)
Known ejection: 383 (86.7% of total)

FARS Data

Restraint Used	Count	Percentage
Yes	944	4.1
No	21,868	95.9

Ejection Status	Count	Percentage
None	22,933	75.8
Total	6,247	20.6
Partial	1,087	3.6

Total occupants: 30,855
Known restraint: 22,812 (73.9% of total)
Known ejection: 30,267 (98.1% of total)

dent scene as rural or urban, the vehicle size, and such occupant factors as seat area, location, sex, and restraint use are consistent. Thus, over all, the NCSS file appears to be a good basis for estimates of accident factors.

Differences between the two files for estimates of proportions of rollovers and ejections appear to be the result of the coding practices and investigative methods. In analyzing these variables, it is necessary to understand under what definitions and conventions the data were collected.

NCSS areas have a slightly higher proportion of younger fatalities. Table 7 shows that the age distributions for fatalities differ between teams because of the areas they serve (many of which are college areas). Also, NCSS has a lower proportion of single-vehicle accidents. These differences, while statistically significant, are small (of the order of 10%) and should not drastically affect the results of the analysis. Despite these differences, NCSS is the best available source of accident data which includes in-depth investigations of physical evidence and official reports.

TABLE 7

NCSS Age by Team:
Counts and Percentages of Known Data
for Fatalities only

Team	Under 30	30 and Over
Calspan (New York)	35 (61%)	22 (39%)
HSRI (Michigan)*	35 (67%)	17 (33%)
Indiana*	61 (62%)	37 (38%)
Kentucky*	35 (70%)	15 (30%)
Miami	9 (43%)	12 (57%)
Southwest (Texas)	87 (61%)	55 (39%)
Dynsc (California)	10 (59%)	7 (41%)
Aggregate	272 (62%)	165 (38%)
FARS	17,422 (57%)	13,294 (43%)

* Indicates a college area.

III.   ESTIMATED PROBABILITY OF FATALITY IN NCSS ACCIDENTS

Method

The probability of fatality was estimated for the NCSS sampling areas by computing the proportion of fatalities for each of several characteristics. Thus, the probability of fatality in NCSS rural accidents is estimated by the weighted data by:

$$\frac{\text{fatalities in rural accidents}}{\text{all occupants in rural accidents}} = \frac{264}{13,157} = 0.02007.$$

It must be emphasized that these estimates are for tow-away passenger car accidents in the NCSS sampling areas only, and no estimates of national fatality rates are made at this time.

The number of fatalities is the actual count of occupants

killed in the NCSS areas. However, other accidents are sampled
at 100 percent, 25 percent, or 10 percent, depending upon the
severity of the resulting injuries. The less severe, but more
frequent, accidents are then weighted by the inverse of the sample fraction (1,4, or 10) to estimate the actual occurrence in
the NCSS areas. This has the effect of increasing the size of
the file, and thus of the apparent statistical significance of
the findings. In estimating the fit of the CONTAB models, no adjustments have been made for this effect.

Fatality by Accident Type

The fatality rate for occupants, given various accident level
descriptors, is shown in Table 8. The number of vehicles involved
in the accident distinguishes car/car from car/object impacts.
For single-vehicle accidents, which are vehicles hitting an object
or the ground, the estimated probability of fatality is over twice
as high as for multi-vehicle impacts. Single-vehicle accidents
account for 41 percent of the fatalities, but only 24 percent of
the total occupants in NCSS crashes. Thus, for towed NCSS area
vehicles, striking something other than a vehicle has a higher
fatality rate than does striking another car.

The estimated probability of fatality, given a rural or an urban land use, is also shown in Table 8. Rural accidents have
over 5 times as high a risk as do urban accidents. While under
23 percent of the occupants were involved in rural accidents,
these people represented almost 60 percent of the fatalities.
Apparently, urban impacts are relatively lower-velocity, less
severe crashes than are those in the less restricted rural environment. This does not imply that cities are safer places to
drive. No exposure data is available which could measure the
amount of driving done, in miles or in hours. These probabilities
of fatalities are conditional on an accident occurring and no estimates of the probability of the imapct can be made at this time.

Also estimated in the same table is the frequency of fire involvement in the accident. This variable does not record which
vehicle suffered fire damage, but only that at least one of the
vehicles did. The fatality rate in the presence of fire was 19
times the fatality rate in non-fire accidents. In part, this
probably is an indication of the severity of the crash—more
severe impacts produce both fires and fatalities. It cannot be
assumed that the fire caused the fatality.

Table 9 relates the probability of fatality to a description
of the accident. Accident type is a finer breakdown of the number
of vehicles involved which is shown in Table 8. One-vehicle accidents include car/fixed object categories and principal rollovers. Two-vehicle accidents include car/vehicle categories and
sideswipes. Three or more vehicles in an accident are generally
a chain collision. Because a single description of the whole accident is required, this variable is not as precise as the vehicle-

TABLE 8

NCSS Estimated Probability of Fatality
by Various Accident Variables

Number of Vehicles	Fatalities	All Occupants	Estimated Probability of Fatality
One	182	13,840	0.01315
Two	232	37,152	0.00624
Three or more	28	7,077	0.00396
Rural/Urban			
Rural	264	13,157	0.02007
Urban	178	44,865	0.00397
Fire			
Yes	30	222	0.13514
No	412	57,847	0.00712

level variables in describing the damage to each vehicle, but it does give an approximation of the accident sequence.

Car/fixed object accidents have twice the fatality rate that occupants of car/vehicle collisions suffer. Comparisons within the car/vehicle categories reveal that the risk in a head-on crash is two and a half times that in a side impact. (Fatality rates for a side impact accident include the occupants of both the car struck in the side and the striking vehicle). Angle impacts have lower rates than the corresponding direct hits in both the case of frontal and side collisions. In the case of impact with a fixed object, a towed side impacted vehicle has three times the fatality rate of a towed frontally damaged car.

The highest fatality rates are for accidents described as side into a fixed object, rollover, and head-on. The lowest rates are for sideswipes, chain collisions, and rear impacts.

Fatality by Vehicle Size and Damage

Table 10 shows fatality rates by various sizes of vehicles, as defined previously by the length of the wheelbase. While "minicar", the smallest size, has the highest fatality rate, and "large", the largest size, has the lowest fatality rate, it is not clear if there is a steady trend of increasing rate with decreasing size. The average of the three smallest categories is 0.00878 as constrasted with 0.00756 for the three largest categories, indicating that size may be an advantage for occupants of towed vehicles.

TABLE 9

*NCSS Estimated Probability of Fatality
by Accident Type*

Type of Impact	Fatalities	All Occupants	Estimated Probability of Fatality
Car/vehicle:	220	34,729	0.00633
Head on	83	5,135	0.01616
Angle front	4	910	0.00440
Side	106	16,809	0.00631
Angle side	13	3,789	0.00343
Rear	14	8,086	0.00173
Car/fixed object:	127	10,073	0.01261
Front	70	7,942	0.00881
Side	56	2,033	0.02755
Rear	1	98	0.01020
Other:	87	11,789	0.00738
Principal rollover	58	2,150	0.02698
Sideswipe	-	1,120	-
Undercarriage	2	721	0.00277
Chain collision	1	2,674	0.00037
Other/unknown	26	5,124	0.00507

(Note: Since this variable describes the entire accident, car/vehicle accident types include counts of all
occupants in all involved vehicles despite individual
damage areas).

Estimates of occupant fatality rates for various vehicle damage descriptions are also shown in Table 10. Rollovers appear to
have the highest risk of fatality for NCSS tow-away accidents: four
times the fatality rate of non-rollovers in general. Note that
while only 5 percent of all occupants were in a vehicle which
rolled over, these people are 18 percent of all fatalities. A
comparison of the specific impact areas for non-rollovers shows a
higher fatality rate for vehicles hit in the side than for frontally damaged vehicles.
A closer examination of the typical side impact is useful because of the wide range of accidents which may result in a particular damage area. Shown in Table 11 is fatality information for
two-vehicle accidents in which one passenger car received frontal
damage as the result of striking another car in the side. As a
control, only accidents in which both vehicles required towing
are considered. Note that in this case the occupants of the

TABLE 10

NCSS Estimated Probability of Fatality
by Vehicle Size and Damage

Size	Fatalities	All Occupants	Estimated Probability of Fatality
Minicar	6	536	0.01119
Subcompact	34	4,457	0.00763
Compact	37	3,774	0.00980
Intermediate	82	10,527	0.00779
Full Size	47	6,157	0.00763
Large	19	2,899	0.00655

Vehicle Damage	Fatalities	All Occupants	Estimated Probability of Fatality
Rollover	75	2,337	0.03209
Non-rollover:	336	44,458	0.00756
Back	2	2,707	0.00074
Front	184	28,163	0.00653
Left	73	6,539	0.01116
Right	75	6,302	0.01190
Other known	2	747	0.00268

struck vehicle have over eight times the fatality rate of the occupants of the striking vehicle.

The NCSS file is, for practical reasons of accident identification and standardization, limited to accidents in which at least one vehicle was damaged sufficiently to require towing. Therefore, it is important to consider the question of whether the striking or the struck vehicle is more often towed. A towed vehicle is more often the case vehicle for which occupant fatality information is available. A large difference in the rate of towing would mean that occupant information for the less-often towed vehicle would be biased towards the more severe accidents which resulted in towing.

Table 11 shows that the rate of towing in side impact accidents is almost the same for the striking and struck vehicle. The struck car is towed 74 percent of the time, the striking car is towed 75 percent of the time, and in 49 percent of the cases both vehicles are towed. It appears that the towing criterion does not bias the file towards either side or front damaged vehicles, and the relative fatality rates should be consistent with the total accident picture of the NCSS sampling areas for this particular type of crash.

TABLE 11

*NCSS Estimated Probability of Fatality
and Towing for Side Impacts*

*Two-vehicle accidents, one car striking another in the
side, and both vehicles sufficiently damaged to require
towing:*

Vehicle Damage	Fatalities	All Occupants	Estimated Probability of Fatality
Front (striking)	5	4,617	0.00108
Side (struck)	40	4,428	0.00903
Left side	14	2,124	0.00659
Right side	26	2,304	0.01128

*Two-vehicle accidents, one car striking another in the
side: rate of towing required:*

Side	Front Towed	Front Not Towed	Total
Towed	2,947 (49.4%)	1,486 (24.9%)	4,433 (74.3%)
Not Towed	1,537 (25.7%)	-	1,537 (25,7%)
Total	4,484 (75.1%)	1,486 (24.9%)	5,970 (100 %)

The direction of force (DOF) for horizontal impacts is measured
from 1 to 12 and corresponds to the hours on a clock. A DOF of 12
is perpendicular to the front of the vehicle. Table 12 gives fatality rates for the various force directions, including zero for non-
horizontal forces such as an undercarriage. The DOF is not a substitute for the area of damage, but the two variables together
give information on the impact. A DOF of 1 to the front is slightly less than a direct impact, while a DOF of 1 to the right side
might be a sideswipe. Forces from the clock positions 2-4 and 8-10
are often side damage, while 11-12-1 and 5-7 are frequently front
and rear damage, respectively. Thus, the fatality rate for each
DOF is similar to that for the particular damage area which is often associated with it.
    In addition, fatality rates for the DOF categories which corre-
spond to direct impacts into one of the four surfaces are higher
than for indirect or glancing impacts. O'clock positions 3,6,9
and 12 have higher rates than do the adjoining clock positions

TABLE 12

NCSS Estimated Probability of Fatality
by Principal Direction of Force

DOF	Fatalities	All Occupants	Estimated Probability of Fatality
0 (Non-horizontal)	63	2,739	0.02300
1	42	5,561	0.00755
2	35	4,531	0.00772
3	15	694	0.02161
4	10	574	0.01742
5	2	496	0.00403
6	2	2,440	0.00082
7	1	593	0.00169
8	10	747	0.01339
9	11	792	0.01389
10	43	4,102	0.01048
11	40	7,286	0.00549
12	136	16,181	0.00840
11,12,1	218	29,028	0.00751
2,3,4	60	5,799	0.01035
5,6,7	5	3,529	0.00142
8,9,10	64	5,641	0.01135

representing less direct impacts.

Delta V is the change in velocity during impact, and measures the severity of the crash. A high delta V indicates that a large amount of energy was absorbed during impact by the crushing of the vehicles. The CRASH program is an algorithm which is used to estimate the change in velocity, and the values in Table 13 are based upon measurements of the vehicle damage.

Consistently, higher delta V categories have associated with them higher fatality rates. Ranging from only one fatality for 5,939 occupants in the under 7 mph category through one quarter of the group involved in impacts with delta V over 48 mph, there is an increase in the estimated probability of fatality accompanying the increased crash severity.

The CRASH program was not run for almost half of the occupants. The algorithm is not applicable in accidents involving non-horizontal forces, such as rollovers and undercarriage; or when impacting a yielding fixed object, such as is the case when striking and breaking a small tree. In other cases, there is not enough damage information available, as when a vehicle is driven out of the area, or otherwise cannot be located or examined.

If the missing values tended to be for the more severe crashes, this could bias estimates based upon the known factor. The percentages of known delta V for fatalities is almost the same as for the

TABLE 13

NCSS Estimated Probability of Fatality
by Delta V

Delta V	Fatalities	All Occupants	Estimated Probability of Fatality
00-06	1	5,939	0.00017
07-12	8	12,484	0.00064
13-18	17	7,192	0.00236
19-24	28	2,896	0.00967
25-30	42	1,249	0.03363
31-36	39	628	0.06210
37-42	22	328	0.06707
43-48	14	68	0.20588
Over 48	55	220	0.25000
Known	226	31,004	0.00729
Unknown	216	27.065	0.00798
00-12	9	18,423	0.00049
Over 12	217	12,581	0.01725

Percentages of known delta V:

Fatalities: 51.1%
All Occupants: 53.4%

file as a whole: 53 percent vs 51 percent. The fatality rate for occupants with known delta V is slightly lower than that for occupants with unknown delta V: 0.00729 va 0.00798. This difference is small enough to not be considered significant at the 0.05 chi-square level. Using this criterion, the occupant fatality rate for known delta V is consistent with that for the file as a whole. Thus, the large number of cases with unknown delta V does not seem to bias the results significantly in favor of, or against, the less severe crashes.

The delta V is estimated to be correct to within 20 percent, for those crashes for which it was designed. Thus, at the higher values of delta V, the error becomes a larger factor in miles per hour. A delta V of 20 mph is estimated to be in error by about ±4 mph. However, a computed delta V of 40 mph, can easily be in error by ±8 mph. This error is introduced by the many simplifying assumptions about the accident sequence crash dynamics and vehicle stiffness characteristics which are necessary in an algorithm such as delta V. The reader should consider these estimates with caution, and not attempt to equate them to barrier equivalent velocities.

TABLE 14

*NCSS Estimated Probability of Fatality
by Seat Location*

Fatalities

Location	Seat Area			Total
	Left	Middle	Right	
Front	292	10	91	394
Second	11	6	18	36
Third	–	1	–	1
Total	303	17	109	

All Occupants

Location	Seat Area			Total
	Left	Middle	Right	
Front	36,990	2,066	12,059	51,236
Second	2,214	1,148	2,532	6,210
Third	2	2	1	57
Total	39,207	3,216	14,593	

Ratio (Fatalities/All Occupants)

Location	Seat Area			Total
	Left	Middle	Right	
Front	0.00789	0.00484	0.00755	0.00769
Second	0.00497	0.00523	0.00711	0.00580
Total	0.00773	0.00529	0.00747	

Fatality by Occupant Factors

Table 14 presents seating information for fatalities, all occupants, and the fatality rate. The total shown for a row or column may be larger than the sum of the elements because of unknown
data, or categories not shown. For example, the number of people
killed while lying across the third seat would be included in the
total for the third seat location, but not in any of the associated
seat areas shown.

Sitting in the second seat location and/or in the middle seat
area is associated with a lower fatality rate. The driver's seat
has the highest estimated probability of fatality, and 64 percent
of the total occupants were driving when the accident occurred.

TABLE 15

*NCSS Estimated Probability of Fatality
by Age and Sex*

Fatalities

| Age | Sex | | Total |
	Male	Female	
Under 20	87	39	126
20-29	104	42	146
30-39	46	11	57
Over 39	66	42	108
Total	305	136	

All Occupants

| Age | Sex | | Total |
	Male	Female	
Under 20	10,349	7,310	17,734
20-29	11,423	7,102	18,556
30-29	3,959	3,121	7,120
Over 39	7,285	6,014	13,323
Total	33,558	23,765	

Ratio (Fatalities/All Occupants)

| Age | Sex | | Total |
	Male	Female	
Under 20	0.00841	0.00534	0.00710
20-29	0.00910	0.00591	0.00787
30-39	0.01162	0.00352	0.00801
Over 39	0.00906	0.00698	0.00811
Total	0.00909	0.00572	

The third seat location does not have enough data for a meaningful fatality rate.

Fatality data by age and sex are summarized in Table 15. The unknown rates for these variables are very low: both age and sex are known for approximately 99 percent of the file. For those few cases where, for example, the age of a male is not known, the occupant is included only in the column total for males. Therefore the totals are sometimes slightly higher than the sum of the categories of the variable shown.

While females account for 41 percent of the total occupants, only 31 percent of the fatalities are female. This is reflected in the fatality rates by sex. Males have almost 1.6 times the fatality rate of females.

The totals across sex of the four age categories show a steady increase in the fatality rate with the increase in age. Within each age category, females have the lower fatality rate. A multiplicative model can be used to quantify the separate effects of age and sex on the fatality rate, and perhaps provide insight into the probability of fatality as estimated by the data. This is done in a later section.

The fact that seating position, age, and sex appear related to the fatality rate helps to explain the wide range of injuries suffered by different people in similar accident situations, and by the various occupants of a single accident. However, it does not immediately suggest means of reducing the risk of fatality or injury.

Restraint use is one variable which is almost entirely the result of conscious choice on the part of the individual. Most cars are equipped with a restraint system and most people are physically able to wear them. However, of all occupants on the NCSS file whose restraint use was known, only 8 percent were using any kind of restraint. For fatalities, only 5 percent were restrained. The data are shown in Table 16.

The estimated probability of fatality for increasingly restrained categories (no restraint, lap only, lap and torso) decreases steadily (from 0.005 to 0.004 to 0.002). An estimate of the percent reduction in the rate resulting from restraint effectiveness can be computed as follows for the fatality rates of the restraint system:

Effectiveness = percent reduction in fatality rate

$$= \frac{\left(\begin{matrix} \text{Rate with} \\ \text{no restraint} \end{matrix}\right) - \left(\begin{matrix} \text{Rate with} \\ \text{restraint} \end{matrix}\right)}{(\text{Rate with no restraint})} \times 100.$$

Using this formula, unadjusted for any other differences that may exist between the various occupant groups, results in the following estimates of belt effectiveness:

Lap only: $\dfrac{0.00545 - 0.00421}{0.00545} = 23\%$,

Lap and torso: $\dfrac{0.00545 - 0.00236}{0.00545} = 57\%$.

These estimates are necessarily crude because of the small number of restrained occupants, but do indicate that restraints are effective in reducing the death rate.

The issue of restraint use is related to that of ejection.

TABLE 16

NCSS Estimated Probability of Fatality by Restraint Use, Ejection, and Entrapment

Restraint Used	Fatality	All Occupants	Estimated Probability of Fatality
None	247	45,297	0.00545
Lap & Torso	5	2,117	0.00236
Lap only	8	1,898	0.00421
Other known	-	140	-
Unknown	50	8,617	0.00580

Percentages of Known Restraint Use:

Fatalities: 79.8%
All Occupants: 81.0%

Ejection	Fatality	All Occupants	Estimated Probability of Fatality
None	273	56,911	0.00480
Total	93	429	0.21678
Partial	17	125	0.13600
Unknown	59	604	0.09768

Ejection	Fatality	All Occupants	Estimated Probability of Fatality
No	267	56,926	0.00469
Yes	79	404	0.19554
Unknown	96	739	0.12991

Percentages of Known Ejection/Entrapment

Fatalities: 86.7%
All occupants: 98.7%

Table 16 shows the higher fatality rates associated with ejection types, especially total ejection. Partial ejectees have 28 times the fatality rate of non-ejectees. Total ejectees have an even higher rate — 45 times that of non-ejectees. Fully one-fifth of ejectees are killed in the accident. While only 1 percent of all occupants are ejected to some degree, these people represent almost 29 percent of the fatalities.

For those occupants about whom entrapment information is available, only 0.7 percent were trapped in the vehicle, but these are almost 26 percent of all fatalities. Entrapment does not include doors which are locked by damage, but only cases where a part of the occupant's body is physically restrained by the vehicle. Trapped occupants have almost 42 times the fatality rate of non-trapped occupants.

Ejection and entrapment are, at least in part, measures of the severity of the accident and the amount of energy absorbed in the crash. Also, the high fatality rate for ejectees is consistent with that for rollovers, because ejection is often a result of a rollover. Table 17 separates ejection status by rollover and non-rollover accidents.

Note the difference for the two types of accidents. In both cases, ejections coincide with a higher rate of fatality, but the relative risk of partial ejection vs total ejection is reversed. For non-rollovers, the fatality rate of partial ejectees is almost 18 times that of non-ejectees; for total ejectees, the rate is almost 44 times that of non-ejectees. However, for rollovers it is partial ejections which have the highest fatality rate. Total ejectees have over 20 times the rate for non-ejectees, but for partial ejectees vs non-ejectees, the rate is 30 times as high. Possibly, when a vehicle is rolling, it is less dangerous to be thrown clear than to be partially ejected and risk being crushed as the vehicle continues rolling.

The table also shows that almost 13 times as many occupants are ejected in rollovers as in non-rollovers. Of 1,914 rollover occupants with known ejection status, 8.9% were ejected. For non-rollovers, 0.8% of occupants with known ejection status were ejected.

The figures in Table 18 are presented to indicate the extent to which occupant descriptors affect a particular person's risk of fatality. The Abbreviated Injury Scale (AIS) is a standardized dictionary of injuries and associated severity levels. The severity ranges from 0 (no injury) through 6 (currently untreatable). Typical examples of the rating are included in Table 21 as illustrations.

The Overall AIS (OAIS) assesses the net effect of the combined injuries, and is generally equal to the highest individual injury sustained. The scale does not reflect such considerations as occupant age or general health. An amputated arm receives a rating of 4, despite an individual's ability to recover from such an injury.

The injuries sustained ranged from one person who died as the result of minor injuries (OAIS of 1) through the 88 who died from injuries which are currently untreatable (OAIS of 6). Death is the result of the combination of the specific injury and the individual's reaction to it.

In over half of the cases, there were no injuries or injury levels recorded. These people are rated as an OAIS of 8 (injury severity unknown) because they were not examined by a physician. In areas where the coroner is not a medical doctor, no medical report

TABLE 17

NCSS Estimated Probability of Fatality
by Rollover and Ejection Status

Description	Fatality	All Occupants	Estimated Probability of Fatality
Rollover	75	2,337	0.03209
Ejection Type			
None	25	2,040	0.01225
Partial	9	24	0.37500
Total	37	147	0.25170
Unknown	4	126	0.03175
Non-Rollover	367	55,732	0.00659
Ejection Type			
None	248	54,871	0.00452
Partial	8	101	0.07921
Total	56	282	0.19858
Unknown	55	478	0.11506

Percentages of Known Ejection Status:

Rollovers:	Fatalities:	94.7%
	All Occupants:	94.6%
Non-Rollovers:	Fatalities:	85.0%
	All Occupants:	99.1%

is filled out. Thus, information on the nature of the injuries
and the associated contact points is not available on the NCSS computer file.

Further information on injury coding is available in a handbook,
The Abbreviated Injury Scale (AIS), developed as a joint effort of
the American Medical Association, The Society of Automotive Engin-
eers, and the American Association for Automotive Medicine. (Reference 6).

Fatality Rate Summary

At the accident level, several variables had different fatality rates associated with them. Single-vehicle accidents had
higher rates; in particular, side collisions into fixed objects
and rollovers. Of multi-vehicle accidents, head-on collisions had
the highest proportion of fatally injured occupants. Rural accidents and those involving fire also had higher than average fatality rates.

TABLE 18

NCSS Probability of Fatality
by Overall AIS

Overall AIS	Fatalities	All Occupants	Estimated Probability of Fatality
0 (no injury)	-	27,622	-
1 (minor)	1	7,999	0.00013
2 (moderate)	-	1,631	-
3 (severe)	4	823	0.00486
4 (serious)	17	209	0.08134
5 (critical)	92	168	0.54762
6 (maximum)	88	88	1.00000
8 (inj. sev. unk.)	235	9,163	0.02565
9 (unk. if inj.)	-	9.956	-

Percentages of Known Overall AIS:

Fatalities: 46.8%
All Occupants: 66.4%

AIS Level	Typical Injury
1	arm - superficial laceration
2	arm - deep laceration
3	wrist - dislocation
4	spleen - rupture
5	kidney - rupture
6	head - decapitation

Vehicle size appears to be an advantage in accidents, but quantifying this effect is complicated by the towing criterion and the differences in the vehicle reaction to a crash. In a two-car, side impact with both vehicles requiring towing, occupants of the striking car appear to have the advantage. Direct impacts, as determined by the direction of force, and high impact speeds, as measured by delta V, result in higher than average fatality rates.

Several occupant characteristics also appear to increase the risk of fatality. In particular, front or outside (left of right) seating, older age categories, and males are associated with higher fatality rates. Restraint use, non-ejection, and non-entrapment are associated with lower fatality rates. Fatalities are the result of a wide range of injuries, as indicated by the Overall AIS.

IV. CONTINGENCY TABLE ANALYSES

It is sometimes possible to sort out the effects of two inter-acting variables on the fatality rate. This was suggested by four pairs of variables given in the previous section: seat area by lo-cation, age by sex, sex by seating area, and rollover by ejection. In each case, both factors were affecting the fatality rate, but it was difficult to quantify the separate effect of each. A sim-plified model of the NCSS data produced by the APL version of the CONTAB algorithm produces a good fit of fatality by each of these pairs of factors.

CONTAB is a tool of contingency table analysis. It is an iter-ative program which fits an idealized model of the data under such assumptions of independence as are specified by the analyst, and then measures the closeness of the fit. For each of the three cases in the following section, the assumption was made that each of the two factors in the model was independent of the other in its effect on the fatality rate. The model associates changes in the levels of the variable factor with changes in the risk of fa-tality. One possible interpretation of the model is that these changes in the factors cause changes in risk. However, this hy-pothesis cannot be accepted without careful investigation of other evidence and knowledge of the physical situation in each case.

The next three sections describe the results of fitting models to the NCSS data and give additional explanation of the modeling process.

The Information in Contingency Tables by D. Gokhale and S. Kull-back is a useful reference for the concepts involved in this pro-cess, as well as for the practical aspects of using a CONTAB com-puter program (Reference 7).

Model of Seat Area and Location

In an attempt to separate the effect of the set area from that of the location, the CONTAB program was used to fit a model to the data. A model which fits the data will produce estimates of the effects of each variable. In this case, it means that an estimate of the increased risk associated with the front seat can be made independently of the effect of left vs middle vs right seat.

Table 19 shows the result of the model fitting under the assump-tion that the two variables act independently to determine the fa-tality rate. The statistics at the top of the page measure the goodness-of-fit of the model to the data. The Information Statis-tic (I.S.) and the Degrees of Freedom (D.F.) determine the probab-ility (P) of two independent variables differing as much as do these data from the model produced. This model produces a fit close enough that half of all independent relationships would dif-fer this much from a similar model. No adjustments in the probab-ility have been made for the effect of the weighting factor on the sample.

TABLE 19

NCSS Data and CONTAB Fitted Model of Fatality
by Seat Location:
I.S. = 1.37; *D.F.* = 2; *P* = 0.50

Fatalities

Location	Seat Area		
	Left	Middle	Right
Front	292	10	91
	(288.9)	(11.0)	(93.1)
Second	11	6	18
	(14.1)	(5.0)	(15.9)

Non-Fatalities

Location	Seat Area		
	Left	Middle	Right
Front	36,698	2,056	11,968
	(36,701.1)	(2,055.0)	(11,965.9)
Second	2,203	1,146	2,514
	(2,199.9)	(1,143.0)	(2,516.1)

Ratio from Model: Fatalities/Non-Fatalities

Location	Seat Area		
	Left	Middle	Right
Front	0.0079	0.0054	0.0078
Second	0.0064	0.0044	0.0063

Inferences from Model

The Odds of Fatality are:

1.2 times as high for front as for second seat occupants;

1.5 times as high for left or right as for middle seat occupants.

The actual data and the estimates from the model (in parentheses) are shown for fatalities and non-fatalities. Since the model fits sufficiently well, it is useful to interpret the model. The odds ratio of the model is computed by dividing fatalities by non-fatalities for each seating category. For example, the fatality odds for a driver are $(288.9)/(36,701.1) = 0.0079$. The model estimates the relationships between the odds ratios. For each seat area, the value for the front seat is 1.2 times that for the second seat. For each location (front or second), the value is just slightly higher for the left seat that for the right; and for each, it is 1.5 times that for the middle seat area. Thus the model shows an increased odds of fatality associated with the window seats and front seat, and a quantification of each.

Model of Age and Sex

Table 20 shows the data from the NCSS file and the results of the CONTAB model (in parentheses) under the assumption that age and sex are independent of each other in their effect on the fatality rate. The model fits well enough to be a useful interpretation of the data: approximately 65 percent of all independent relationships would have at least this large a discrepancy between the raw data and the idealized model, just by chance. This is a measure of the confidence in the model as an explanation of the data.

The model implies that the odds of fatality for males is 1.6 times that of females, and that the odds for people under 30 is 1.1 times that for the older group. A female under 30 has the lowest odds ratio, 0.0055; and a male of 30 years or more has the highest odds ratio, 0.0099. This reflects the effects of the higher risk implied by both his sex and his age.

Model of Sex and Seating Location

A confounding aspect of the estimation of the odds of fatality, by either seating location or by sex, is that males are much more likely to be driving during an accident than are females. While 72 percent of the males on the NCSS file were driving when the accident occurred, only 53 percent of the females involved were drivers in the accident.

The model presented in Table 21 attempts to sort out these two effects (sex and whether or not the person was a driver) using tools of contingency table analysis. The better fit of the data resulted when sex was used as a predictor of fatality, but seating location was not. Inclusion of the seating location as a factor did not significantly improve the accuracy of the prediction, and this model was discarded.

Therefore, the estimates from the model shown in the table are that males have 1.6 times the odds ratio of fatality that females have, and that this ratio does not depend on whether or not the person was driving at the time of the accident. It appears that the higher percentages of males who were drivers is not a cause of their higher fatality rate.

TABLE 20

*NCSS Data and CONTAB Fitted Model of Fatality
by Age and Sex:
I.S. = 0.203; D.F. = 1; P = 0.65*

Fatalities

Age	Sex	
	Male	Female
Under 30	191 (193.1)	81 (78.9)
30 and over	112 (109.9)	53 (55.1)

Non-Fatalities

Age	Sex	
	Male	Female
Under 30	21,581 (21,578.9)	14,331 (14,333.1)
30 and over	11,132 (11,134.1)	9,082 (9.079.9)

Ratio from Model: Fatalities/Non-Fatalities

Age	Sex	
	Male	Female
Under 30	0.0089	0.0055
30 and over	0.0099	0.0061

Inferences from Model

The Odds of Fatality are:

1.1 times as high for the older group as
for the younger;

1.6 times as high for males as for females.

Model of Rollover and Ejection

The CONTAB algorithm was used on the ejection/rollover data in
an attempt to separate out the effects of the two variables. Table
22 gives the NCSS data and the value produced by the model (in

TABLE 21

*NCSS Data and CONTAB Fitted Model of Fatality
by Sex and Seating Location:
I.S.* = 3.147; *D.F.* = 2; *P* = 0.21

Fatalities

| Sex | Seating Location | |
	Driver	Non-Driver
Male	228 (219.6)	77 (85.4)
Female	64 (72.1)	72 (63.9)

Non-Fatalities

| Sex | Seating Location | |
	Driver	Non-Driver
Male	23,933 (23,941.4)	9,320 (9.311,6)
Female	12,542 (12,533.9)	11,087 (11,095.1)

Ratio from Model: Fatalities/NonlFatalities

| Sex | Seating Location | |
	Driver	Non-Driver
Male	0.0092	0.0092
Female	0.0058	0.0058

parentheses) using ejection as a yes/no variable. The model fits well enough that 20 percent of all independent relationships would have a larger discrepancy between the data and the model of independence, as measured by the information statistic and the degrees of freedom.

The odds of fatality computed from the values of the model quantify the effects of rollovers and ejections. Ejection increases the odds ratio by a factor of 40; rollover occupants have 2.3 times the odds ratio of non-rollover occupants. Since the model assumes independence between ejection and rollover effects, an ejectee in a rollover has an odds ratio equal to the odds ratio of a non-ejected, non-rollover occupant times the product of the increments of the two variable effects. That is,

$$0.4147 = 0.0046 \times 39.9 \times 2.3.$$

TABLE 22

NCSS Data and CONTAB Fitted Model of Fatality
by Rollover and Ejection:
I.S. = 1,669 ; D.F. = 1 ; P = 0,20

Fatalities

Ejection	Rollover	
	Yes	No
Yes	46 (50.1)	64 (59.9)
No	25 (20.9)	248 (252.1)

Non-Fatalities

Ejection	Rollover	
	Yes	No
Yes	125 (120.8)	319 (323.2)
No	2,015 (2,019.2)	54,623 (54,618.8)

Ratio from Model: Fatalities/Non-Fatalities

Ejection	Rollover	
	Yes	No
Yes	0.4147	0.1853
No	0.0104	0.0046

Inferences from Model

The Odds of Fatality are:

39.9 times as high for ejectees as for non-ejectees;

2,3 times as high for rollover as for non-rollover occupants.

This model does not take into account the difference between total and partial ejections. These effects are too interrelated with the rollover variable, as shown in Table 17, to be able to fit an independence model.

Summary of Models

These models are useful in quantifying the risk of fatality and suggest the relative sizes of these risks:

Model: Seat Area and Location

front seat vs rear seat	1.2
window seat vs middle seat	1.5

Model: Age and Sex

old vs young	1.1
male vs female	1.6

Model: Sex and Seating Location

male vs female	1.6
driver vs other passenger	1.0

Model: Rollover and Ejection

rollover vs non-rollover	2.3
ejection vs non-ejection	39.9

Certain characteristics on the NCSS file are more often associated with fatality than are others, and it is possible to estimate the incremental risk using these models. This is not the same as identifying the cause of the increased risk. The models do not take into account the different driving habits of various age groups, or their tendencies toward certain models, sizes, or ages of cars.

Rollovers and ejection are associated with a much higher fatality rate. Preventing rollover and ejection will have the effect of lowering that rate if people are killed by the vehicle rolling onto them or by severe impacts if they strike the ground. However, to a certain extent rollover and ejection are also indications that the vehicle was moving very fast at the time that the driver lost control, or was struck very hard by another vehicle. These accidents will still have a high risk of fatality even if rollover and ejection are reduced.

In summary, while models are useful in quantifying relationships and separating out complicating factors, they are limited in actual interpretations of results. They are best used as indications of possible productive areas of future study.

V. CONCLUSION

The objective of this study has been to investigate the fatalities of the National Crash Severity Study. To do this, it has seemed desirable to, first of all, put NCSS in the context of the

national experience of traffic deaths by comparing these fatalities
to those on the Fatal Accident Reporting System. The census information available on FARS differs from the NCSS on some key variables:
number of vehicles involved, rollover, age of occupant, and ejection status of occupant. Further analysis of the NCSS file will
determine which of these differences are the result of the choice
of sampling areas. Number of vehicles involved is an important
crash descriptor, and analysis usually considers single and multi-
vehicle accidents separately. Similarly occupant age is an important injury factor, and so is usually controlled for in analysis of
injury severity.

On the other hand, it appears likely that the differences between
the files for rollover and ejection are the results of different,
but equally legitimate coding practices. This means that the defin-
ition and decision rules for these variables are very important to
a determination of their frequency of occurrence in any population.

The second aspect of the study involved estimating relative probabilities of fatality for the NCSS areas. Important factors associated with higher fatality rates were identified and quantified using
fatality rates (as found in the file) and modeling techniques (for
the separation of correlated pairs of factors). For example, higher fatality rates were computed for accidents involving entrapment,
ejection, large changes in velocity during impact, and fire in this
accident. Further analysis is needed to determine which of these
factors are causes of fatality and which are merely indications of
the severity of the collision.

VI. REFERENCES

1. Kahane, C., Smith, R. and Tharp, K.: "The National Crash Severity Study", Report of the Sixth International Technical Conference on Experimental Safety Vehicles. Pub. No. DOT-HS-802-50.
   NHTSA, Washington, D.C., 1976.

2. Hedlund, J.: "A Working Guide to the National Crash Severity
   Study". Unpublished Report, NHTSA, April, 1978.

3. *1977 FARS Annual Report*. NHTSA Publ. No. DOT-HS-8-01954,
   Washington, D.C., November, 1978.

4. Partyka, S.: "Injury Predictors by Area of Damage to the Vehicle". Unpublished Report, NHTSA, July, 1978.

5. Partyka, S.: "Rollovers and Injury on the NCSS File". Unpublished Report, NHTSA, October, 1978.

6. *The Abbreviated Injury Scale (AIS)*, American Association for
   Automotive Medicine, Morton Grove, Illinois, 1976.

7. Gokhale, D. and Kullback, S.: *The Information in Contingency
   Tables*, Marcel Dekker Inc., New York, 1978.

THE SEVERITY OF LARGE TRUCK ACCIDENTS†

James Hedlund

National Highway Traffic Safety Administration,
U.S. Department of Transportation

INTRODUCTION

The safety record of large trucks is a source of increasing concern. Many economic considerations argue that larger trucks are cheaper and more efficient long-distance haulers than smaller trucks. The "energy crisis" of 1973 and the subsequent rise in fuel prices and decrease in speed limits strengthened these arguments. The recent "Commercial Vehicle Post-1980 Goals Study" reviewed trucking industry economics in some detail and recommended that longer, wider, and heavier trucks be not only permitted but encouraged.

The increased use of heavy trucks may have a significant effect on highway safety. State and Federal governments have long felt that large trucks are a hazard to automobile traffic, and that their danger probably increases with their size and weight. As a result, numerous State and Federal laws restrict the length, width, weight, and number of units of trucks operating in various jurisdictions. These laws are often based more on intuition and on the highly visible consequences of a few spectacular accidents than on any comprehensive study of large truck safety. Indeed, some claims have been made that large trucks are safer than the highway population as a whole when compared on the basis of accidents per mile travelled.

This study uses data supplied by the Bureau of Motor Carrier

† This paper originally appeared as a technical note in the Mathematical Analysis Division Technical Note series of the DOT/NHTSA National Center for Statistics and Analysis: NHTSA Technical Note DOT HS-802 332, April 1977. The original may be obtained from the National Technical Information Service, Springfield, Virginia 22161.

206

Safety to examine the severity of accidents between cars and articulated trucks. It is found that the accident location has a far
greater effect on accident severity than does the size or weight of
the truck. An accident is most likely to be severe (as measured
by a fatality to a car occupant) if it occurs on a rural, 2-lane
road. Rural, 4-lane roads have proportionately fewer severe accidents, and residential/business roads have proportionately fewer
yet. Once accidents have been classified by location into these
three categories, accident severity does not vary significantly
with either the size or weight of the truck involved.

Data

The best national data on large truck accidents come from the
Bureau of Motor Carrier Safety (BMCS). The BMCS is responsible for
monitoring the safety record of all interstate motor carriers
(trucks and busses). These carriers are required to report to the
BMCS all accidents involving one of their vehicles which results
in either personal injury or property damage in excess of $2,000.
The reporting form gives a rather complete account of the carrier's
vehicle and its involvement in the accident, but provides only minimal information about other vehicles and people. Minor accidents
are of course not reported, nor are accidents involving only intra-
state carriers. So the BMCS files are not a complete census of
truck accidents, but they do give an accurate description of the
accident record of larger trucks.
In order to study the effect of truck size on safety, both
"size" and "safety" must be defined in a manner compatible with the
BMCS data file. Truck size is measured and regulated in many ways:
overall weight, weight/axle, overall length, unit length, width,
number of axles, and type (such as straight truck, tractor-semitrailer, or double bottom). The BMCS file contains both the weight
(loaded weight, at the time of the accident) and the type of the
trucks involved; since each describes a different characteristic
of the vehicle, each was used in the analysis. Safety was interpreted as the severity of an accident given that an accident occurred (since no mileage travelled date are available with which to
compare the number of accidents occurring to various classes of
trucks). The measure of severity used was the incidence of a fatality to a car occupant in a car-truck accident. Fatalities were
chosen to measure severity since they are of greater public interest, are much easier to determine, and are far more likely to be
reported accurately than are injuries. Car-truck accidents were
selected, and only fatalities to car occupants were considered,
since the safety of car occupants was the object of primary interest. (In all car-truck accidents on the BMCS files only 4.7% of
all fatalities were to truck occupants.) Finally, each car-truck
accident was classified as having either no car occupant fatalities
or at least one such fatality. This classification, instead of a
count of the total number of fatalities in an accident, removed

the effect of a few very severe accidents and provided an initial smoothing of the data. (Overall, there was an average of 1.29 fatalities to car occupants for each fatal accident.)

The BMCS files for 1973 and 1974 were examined. Since large trucks were of primary interest, only articulated truck accidents were considered. (An articulated truck is one with two or more units and a flexible joint, such as a tractor-semitrailer combination.) The file was edited to remove obviously erroneous data: an articulated truck was required to have a weight of at least 13,000 lbs. Each accident was then classified according to each of the following variables:

Year: 1973 or 1974 (2 levels);
District: rural or residential/business (2 levels);
Roadway: 2 lane or at least 4 lanes (2 levels);
Truck type: 3,4,5-axle semitrailers or double bottoms (a trac-
   tor-semitrailer-full trailer combination) (4 levels);
Weight: loaded weight at the time of the accident, measured
   in 10,000 lb intervals (7 levels);
Fatality: is there a fatality to a car occupant in the acci-
   dent or not? (2 levels).

The file yielded 14,574 accidents.

A table of the complete data, classified into 448 categories by the 6 variables listed above is given in Appendix A. A summary, classified by truck type and fatality only, is given in Table 1.

TABLE 1

Summary of BMCS Data

Truck Type	Non-Fatal Accidents	Fatal Accidents	Total Accidents	Odds of Fatality
3 axle	880	76	956	0.086
4 axle	2,155	198	2,353	0.092
5 axle	9,718	1,163	10,881	0.120
double	331	53	384	0.160
total	13,084	1,490	14,574	0.114

Analysis

The data were analyzed to determine which explanatory variables and interactions were needed to predict the severity of a car-truck accident. The precise measurement of severity employed for any category was the odds of fatality: the ratio of the number of accidents with at least one fatality to the number of accidents with no fatalities. (For example, in Table 1, the odds of fatality for

3-axle semitrailers are 76/880 = 0.086.) The algorithm CONTAB, developed at George Washington University, was used to fit log-linear models determined by various combinations of explanatory variables and interactions and to measure the goodness-of-fit of the resulting predicted distributions. A more technical description of the method is given in Appendix B, and the details of the analysis are described in Appendix C.

Overall Results

In the model which describes the data, the odds of fatality to some car occupant involved in a car-truck accident can be predicted accurately from the district (rural, or residential/business) and the roadway (2 lanes, or at least 4 lanes) of the accident. Once this district-roadway combination is known, the odds of fatality do not vary significantly with the type or weight of the truck involved or the year of the accident. More vividly, for car-truck accidents on the BMCS files, it doesn't matter what hit you, only where it hit you. The odds of fatality for each of the four district-roadway categories are given in Table 2.

TABLE 2

Predicted Odds of Fatality for All Truck Accidents

| District | Roadway | |
	2 Lane	4 Lane
Rural	0.211	0.120
Residential/ Business	0.059	0.061

Details of the model, a comparison of the relative effects of the various independent variables, and a discussion of goodness-of-fit may be found in Appendix C.

The simplicity of this model is striking. The model says that for car-truck accidents on rural, 2-lane roads the odds of fatality to some car occupant are 0.211 regardless of the year, the type of truck involved, or the weight of the truck. Thus there is roughly 1 fatal accident for every 5 non-fatal accidents. For rural, 4-lane roads the odds are 0.120, or about 1 fatal for every 8 non-fatals. For residential/business accidents the odds are quite similar regardless of the number of lanes, and are about 1 in 16. Notice the great difference in these odds: rural, 4-lane accidents have twice as high odds as residential/business accidents, and rural, 2-lane accidents have odds $3\frac{1}{2}$ times higher than residential/business accidents.

The conclusions from the analysis are equally striking: accident severity in car-truck accidents can be predicted by the district

and roadway where the accident occurred. Now the three district-
roadway categories are themselves surrogate measures for true caus-
al variables, such as speed and crash type, which are not recorded
on the BCMS file. Rural, 2-lane accidents probably include a higher proportion of high-speed, head-on collisions; rural, 4-lane accidents include relatively more high-speed, front-to-rear or sideswipe crashes; and lower speed crashes of various types probably
predominate in residential/business accidents. An intuitive ranking of these crashes by severity certainly agrees with the ranking
of the model. The important conclusion is not that district and
roadway themselves determine accident severity but that truck type
and weight do not: once accidents have been classified by location
into the three district-roadway categories, accident severity does
not vary significantly with either the size or weight of the truck
involved.

In the following two sections the results for very heavy trucks
and for double bottoms are examined in more detail.

Results for Double Bottom Trucks

Recall from Table 1 that the overall fatality odds for double
bottoms are $53/331 = 0.161$, while for all other semitrailers the
odds are $1,437/12,753 = 0.113$. The difference in the two overall
odds appears to be significant (chi-square = 5.5, significant at
the 0.02 level). However, these overall odds mask the effects of
the district and roadway variables. The odds for double bottoms
and for all other semitrailers in the four district-roadway categories are given in Table 3.

TABLE 3

Predicted Odds of Fatality
for Double Bottom and Semitrailer Accidents

| District | Truck Type and Roadway | | | |
| | Semitrailers | | Double Bottoms | |
	2-Lane	4-Lane	2-Lane	4-Lane
Rural	0.212	0.117	0.191	0.188
Residential/ Business	0.058	0.060	0.119	0.103

From Table 3, the fatality odds for double bottoms are seen to
be larger than the odds for all semitrailers in 3 of the categories
and smaller in 1. However, none of the differences in odds is significant (at the 0.05 level; the rural, 4-lane difference is significant at the 0.1 level).

There are several reasons for the apparent paradox that the

overall odds differ significantly while the odds for individual
district-roadway categories do not.

(1) Double bottoms have a greater proportion of their accidents
    on rural roads than do other semitrailers: 65% compared
to 51%. Rural accidents are more severe than residential/
business accidents, regardless of the type of truck involved.
Most of the observed difference in overall odds is due to this
greater proportion of rural accidents for double bottoms. If
the 384 observed double bottom accidents are distributed among
the 4 district-roadway categories in the same proportions as
the other semitrailer accidents, and if the odds of fatality
for double bottom accidents remain at the observed values,
then the overall odds of fatality for double bottoms would
be only 0.129 (instead of the observed 0.160). The rate of
0.129 is not significantly different (at the 0.05 level) from
the rate of 0.113 observed for semitrailers.

(2) The number of observed double bottom accidents is quite
    small, particularly in residential/business areas. The
odds of fatality must differ from the odds for semitrailers
by a relatively large amount before the difference can be
called significant. In particular, there are only 13 fatal
double bottom accidents in residential/business areas. A decrease of only 5 fatal accidents, from 13 to 8, would be sufficient to lower the double bottom odds for these areas to
those of all semitrailers.

Conclusions: most of the apparent difference in severity of
double bottom truck accidents is due to the greater proportion of
rural accidents among all double bottom accidents. After control-
ling for the district-roadway combination category, the remaining
effects of truck type (double bottoms compared to all other semitrailers) are quite small. The data do not support a conclusion
that accident severity is greater in double bottom accidents.

Results for Very Heavy Trucks

For the purposes of this analysis, a very heavy truck is defined as any articulated truck with a loaded weight exceeding
70,000 lbs. The overall odds of fatality for very heavy truck accidents are 32/2,708 = 0.137, while the odds for other heavy trucks
(all trucks weighing less than 70,000 lbs) are 1,118/10,376 = 0.108.
Again the difference in odds appears to be highly significant (chi-
square = 14.6, significant at the 0.001 level). But, again, this
difference is largely due to the hidden variables of district and
roadway. The odds for very heavy and other heavy trucks in the
district-roadway categories are given in Table 4.
From Table 4, the fatality odds for very heavy trucks are seen
to exceed those for other heavy trucks in each of the 4 categories.
However, only the residential/business, 4-lane difference is signif-

TABLE 4

Predicted Odds of Fatality
for Very Heavy and Other Heavy Truck Accidents

District	Truck Weight and Roadway			
	Other Heavy Trucks		Very Heavy Trucks	
	2-Lane	4-Lane	2-Lane	4-Lane
Rural	0.208	0.118	0.222	0.124
Residential/ Business	0.056	0.054	0.077	0.090

icant at the 0.05 level. For the two rural classes, in which the
highest fatality odds for either weight class are found, the odds
for both weight classes are extremely close. Very heavy trucks
had a greater proportion of their accidents on rural roads than
did other heavy trucks: 58% as compared to 50%. As before, if the
3,080 very heavy truck accidents are distributed among the 4 district-roadway categories in the same proportions as the other heavy
truck accidents, and if the odds of fatality for very heavy truck
accidents remain at the observed values, then the overall odds of
fatality for very heavy trucks would be only 0.129 (instead of the
observed 0.137). Thus about one-third of the observed difference
in overall odds between very heavy and other heavy trucks results
from the differing proportions of accidents on rural roads.

A graph of fatality odds as a function of weight provides greater insight into the explanatory power of weight and of district and
roadway. These fatality odds are plotted in Figure 1 for each of
the 4 district-roadway combinations and for all categories combined.
Note that very heavy trucks, as defined above, constitute the last
data point of each graph (at 75,000 lbs), while other heavy trucks
constitute the first six data points.

The graphs of Figure 1 show clearly the effects described previously.

(1) The overall fatality odds appear to increase with increas-
    ing weight (and the odds for the 75,000 lb weight class
are greater than those of any lesser weight class); but:

(2) the district-roadway classification is far more important
    than the truck weight in determining the fatality odds.
Note that the great separation of the categories: the rural
2-lane odds are quite uniform, and are all higher than any of
the odds for any of the other categories. The rural 4-lane
odds are a bit more varied, still show no definite trend with
increasing weight, and all lie between the lowest rural 2-lane
value and the highest residential/business value. Both resid-
ential/business category odds appear to increase slightly with

212

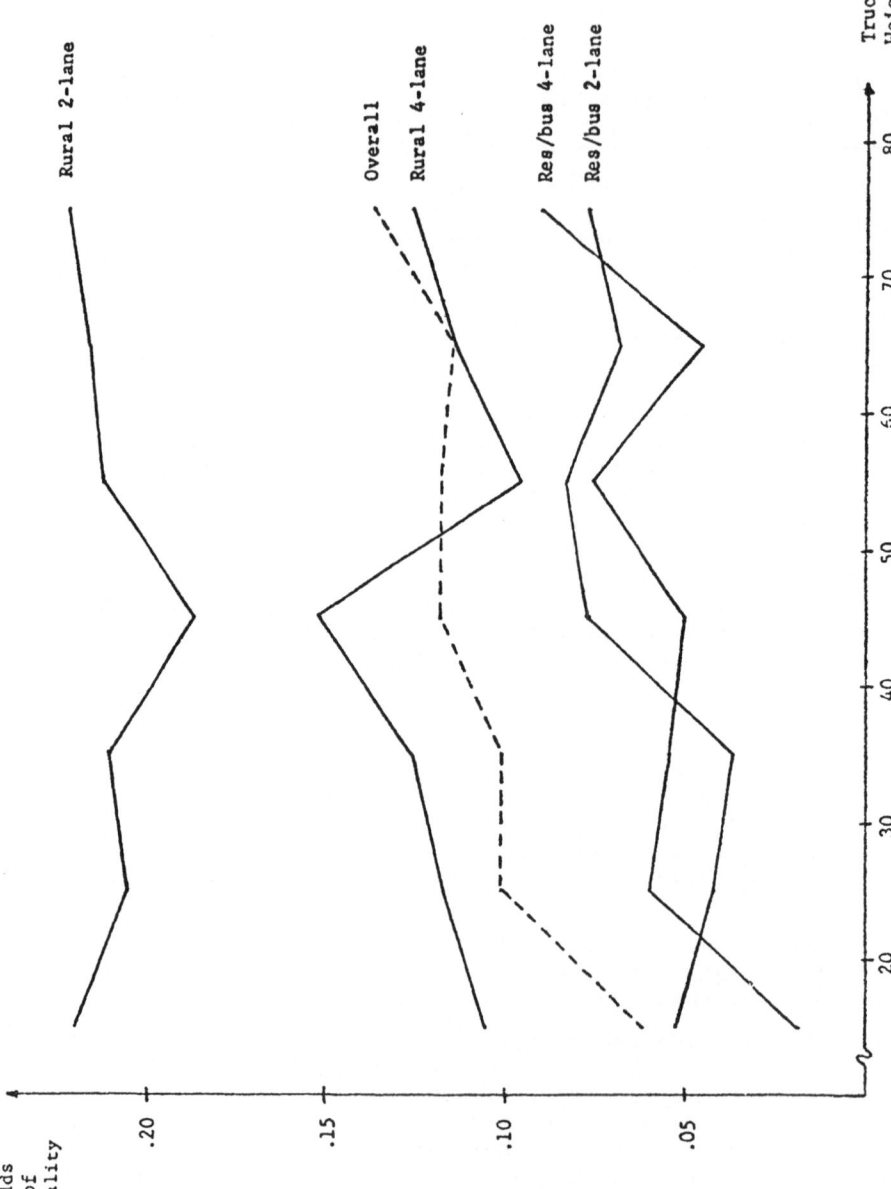

Figure 1.   Odds of fatality and truck weight for different district-roadway conditions.

increasing truck weight, but all the residential/business odds lie below any of the rural odds.

Linear regressions were fit to each of the five sets of odds. The regressions confirm these conclusions: while the overall odds show a highly significant positive slope with increasing weight, neither rural category by itself has a significant slope, and both residential/business categories have moderate positive slopes. Details of the regressions are given in Appendix C.

Conclusions: the effects of truck weight are small compared to the effect of district and roadway. In rural areas, weight appears to have no effect. In residential/business areas there is a slight increase in fatality odds with increasing weight.

APPENDIX A

DATA

    Data for the study are given on the following four pages. Each page contains all observations for one of the four district-roadway combinations. The rows of observations are grouped by truck type, and within truck type by weight. Weight is measured in 1,000 lbs, and the label for each weight class marks the lowest weight of that class. Thus the weight class with the label '10' includes all weights from 1,000 to 19,999 lbs, and the class with the label '70' includes all weights greater than 70,000 lbs. The columns are grouped by the year of the observation. For each year, the three columns record, respectively, the number of accidents with no fatalities to car occupants, the number of accidents with at least one fatality, and the odds of at least one fatality (the ratio of "at least one fatality" accidents to "no fatality" accidents).

DATA: *Rural 2-Lane Accidents*

Fatality		1973			1974		
		No	Yes	Odds	No	Yes	Odds
Weight	10	4	0	0.00	7	2	0.29
	20	26	5	0.19	23	10	0.43
	30	15	4	0.27	23	3	0.13
Three	40	17	1	0.06	13	2	0.15
Axle	50	5	0	0.00	8	1	0.12
Semi-Trailer	60	8	2	0.25	6	1	0.17
	70	12	5	0.42	3	0	0.00
Total		87	17	0.20	83	19	0.23
Weight	10	4	0	0.00	4	1	0.25
	20	93	0	0.22	73	13	0.18
	30	45	10	0.22	33	6	0.18
Four	40	54	10	0.19	30	8	0.27
Axle	50	45	4	0.09	26	6	0.23
Semi-Trailer	60	19	0	0.00	16	3	0.19
	70	17	0	0.00	19	4	0.21
Total		277	44	0.16	201	41	0.20
Weight	10	9	3	0.33	4	1	0.25
	20	292	53	0.18	326	70	0.21
	30	102	19	0.19	126	30	0.24
Five	40	98	21	0.21	98	16	0.16
Axle	50	128	34	0.27	168	34	0.20
Semi-Trailer	60	305	77	0.25	319	66	0.21
	70	378	76	0.20	383	96	0.25
Total		1312	283	0.22	1424	313	0.22
Weight	10	-	-	-	-	-	-
	20	9	2	0.22	5	1	0.20
	30	10	1	0.10	3	2	0.67
Double	40	8	2	0.25	8	1	0.12
Bottom	50	7	2	0.29	5	2	0.40
	60	15	1	0.07	11	1	0.09
	70	20	5	0.25	9	1	0.11
Total		69	13	0.19	41	8	0.20

DATA: *Rural 4-Lane Accidents*

Fatality		No	1973 Yes	Odds	No	1974 Yes	Odds
Weight	10	4	1	0.25	4	0	0.00
	20	12	1	0.08	15	1	0.07
Three	30	9	0	0.00	10	1	0.10
Axle	40	8	1	0.12	9	0	0.00
Semi-Trailer	50	1	1	1.00	2	1	0.50
	60	6	5	0.83	4	1	0.25
	70	3	0	0.00	15	1	0.07
Total		43	9	0.21	59	5	0.08
Weight	10	6	0	0.00	1	0	0.00
	20	56	5	0.09	50	8	0.16
Four	30	51	5	0.10	37	3	0.08
Axle	40	32	4	0.13	36	5	0.14
Semi-Trailer	50	25	3	0.12	23	3	0.13
	60	19	2	0.11	10	3	0.30
	70	13	3	0.23	14	1	0.07
Total		202	22	0.11	171	23	0.13
Weight	10	2	0	0.00	2	1	0.50
	20	189	27	0.14	188	18	0.10
Five	30	91	11	0.12	89	14	0.16
Axle	40	86	17	0.20	105	15	0.14
Semi-Trailer	50	151	9	0.06	135	15	0.11
	60	366	35	0.10	346	39	0.11
	70	304	35	0.15	308	28	0.09
Total		1189	144	0.12	1173	130	0.11
Weight	10	-	-	-	-	-	-
	20	5	0	0.00	5	1	0.20
	30	4	1	0.25	3	2	0.67
Double	40	4	0	0.00	3	1	0.33
Bottom	50	11	1	0.09	8	1	0.12
	60	14	3	0.21	17	1	0.06
	70	15	3	0.20	12	5	0.42
Total		53	8	0.15	48	11	0.23

DATA: *Residential/Business 2-Lane Accidents*

Fatality		No	1973 Yes	Odds	No	1974 Yes	Odds
Weight	10	32	1	0.03	31	2	0.06
	20	65	3	0.05	86	1	0.01
Three	30	36	0	0.00	20	1	0.05
Axle	40	12	0	0.00	8	2	0.25
Semi-Trailer	50	5	2	0.40	7	0	0.00
	60	13	0	0.00	9	0	0.00
	70	1	2	2.00	11	1	0.09
Total		164	8	0.05	172	7	0.04
Weight	10	20	1	0.05	14	0	0.00
	20	119	6	0.05	155	4	0.03
Four	30	77	2	0.03	81	6	0.07
Axle	40	37	6	0.16	56	2	0.04
Semi-Trailer	50	35	3	0.09	45	6	0.13
	60	27	2	0.07	20	2	0.10
	70	14	1	0.07	15	0	0.00
Total		329	21	0.06	386	20	0.05
Weight	10	11	2	0.18	7	0	0.00
	20	296	8	0.03	313	20	0.06
Five	30	105	4	0.04	117	2	0.02
Axle	40	100	5	0.05	78	7	0.09
Semi-Trailer	50	118	11	0.09	123	6	0.05
	60	222	18	0.08	258	16	0.06
	70	250	18	0.07	270	21	0.08
Total		1102	66	0.06	1166	72	0.06
Weight	10	-	-	-	-	-	-
	20	4	0	0.00	1	2	2.00
	30	1	1	1.00	1	0	0.00
Double	40	3	0	0.00	6	1	0.17
Bottom	50	3	0	0.00	1	0	0.00
	60	7	0	0.00	3	0	0.00
	70	6	1	0.17	6	0	0.00
Total		24	2	0.08	18	3	0.17

DATA: *Residential/Business 4-Lane Accidents*

Fatality		1973			1974		
		No	Yes	Odds	No	Yes	Odds
Weight	10	31	1	0.03	29	1	0.03
	20	49	2	0.04	59	3	0.05
Three	30	24	1	0.04	18	1	0.06
Axle	40	8	0	0.00	6	0	0.00
Semi-Trailer	50	4	0	0.00	13	1	0.08
	60	16	0	0.00	4	1	0.25
	70	7	0	0.00	4	0	0.00
Total		139	4	0.03	133	7	0.05
Weight	10	13	0	0.00	16	0	0.00
	20	90	6	0.07	124	3	0.02
Four	30	48	3	0.06	73	2	0.03
Axle	40	34	3	0.09	40	1	0.02
Semi-Trailer	50	33	3	0.09	31	1	0.03
	60	17	1	0.06	31	1	0.03
	70	20	2	0.10	19	1	0.05
Total		255	18	0.07	334	9	0.03
Weight	10	11	0	0.00	11	0	0.00
	20	269	13	0.05	265	21	0.08
Five	30	108	5	0.05	97	8	0.08
Axle	40	93	6	0.06	93	3	0.03
Semi-Trailer	50	149	11	0.07	158	13	0.08
	60	281	14	0.05	278	10	0.04
	70	241	20	0.08	298	31	0.10
Total		1152	69	0.06	1200	86	0.07
Weight	10	-	-	-	-	-	-
	20	7	2	0.28	4	1	0.25
	30	2	0	0.00	2	0	0.00
Double	40	7	0	0.00	0	1	
Bottom	50	6	1	0.17	7	0	0.00
	60	10	0	0.00	12	2	0.17
	70	10	1	0.10	11	0	0.00
Total		42	4	0.10	36	4	0.11

APPENDIX B

ANALYSIS METHODOLOGY

The analysis methodology used is one of a number of methods generally described as contingency table, or discrete multivariate, analysis, These methods have received considerable attention in the recent statistical literature, and a number of specific algorithms for implementing them have been developed. The recent book of Bishop, Fienberg and Holland [1] gives an excellent overall account of the field and provides extensive references to the literature. The algorithm CONTAB used in the present study was developed at George Washington University under the direction of Solomon Kullback. Documentation of the algorithm and examples of its application may be found in [2,3 and 4].

Contingency table methods are appropriate for data that has been cross-classified by a number of variables of interest. Each variable is required to be categorical, so that each observed data point can be placed into exactly one of the finite number of categories of each variable. The restriction to categorical data is not prohibitive, since any data measured on a continuous scale (such as truck weight) can easily be categorized (by dividing the range of weights into 10,000 lb intervals, for example). When all variables are considered simultaneously, they determine a cross-classification into cells. The number M of cells is equal to the product of the number of categories in each variable: if there are 3 variables of interest, with 3,2 and 5 categories, respectively, then the cross-classification defined by them has $M = 3.2.5 = 30$ cells.

The goal of contingency table analysis is to describe the observed data as simply as possible. The data are assumed to obey an underlying multinomial distribution over all M cells. To each cell is associated the probability that a data point selected at random will be classified in that cell. A complete description of the multinomial distribution would require a knowledge of all M probabilities for all M cells. But fewer than M probabilities may suffice to approximate the distribution reasonably well. In fact, an approximation employing fewer probabilities may be more useful than a complete description of the distribution, for the mass of small effects in the complete distribution may mask significant features of the distribution. Contingency table analysis examines various structural models and measures how well the models fit the observed data. The results of the analysis are a model which attempts to maximize both structural simplicity and goodness-of-fit, estimated counts determined by the model for all data cells, an overall measure of goodness-of-fit for the model, an analysis of which variables and interactions are present in the model and which are not, and a measure of the relative importance of those variables and interactions used.

Log-Linear Structure

The analysis begins with a log-linear model for the underlying multinomial distribution: the logarithms of the cell probabilities are expressed as linear functions of the main effects and interactions of the variables. For purposes of illustration, consider a data set with 3 variables, I, J and K. Suppose that variable I has 2 categories (identified as $i = 1$ or 2), variable J has 3 ($j = 1, 2$ or 3), and variable K has 2 ($k = 1$ or 2). Thus the total number of cells is $M = 2.3.2 = 12$. Let $p(i,j,k)$ be the underlying probability that a data point selected at random will be classified in cell (i,j,k). The log-linear model represents each $p(i,j,k)$ in the following form:

$$\log p(i,j,k) = \mu + \alpha_i^I + \alpha_j^J + \alpha_k^K + \alpha_{ij}^{IJ} + \alpha_{ik}^{IK} + \alpha_{jk}^{JK} + \alpha_{ijk}^{IJK} ,$$

where $i = 1,2$; $j = 1,2,3$; $k = 1,2$; and log is the natural logarithm. The 7 terms have the following interpretations:

μ is a general mean,

α_i^I measures the effect of the variable I alone on the probability $p(i,j,k)$. Since these terms measure deviations from the general mean, they must sum to 0 over all values of i: $\alpha_1^I + \alpha_2^I = 0$. Thus there is only one independent parameter α_i^I. The α_j^J and α_k^K measure similar main effects for the J and K variables.

α_{ij}^{IJ} measures the interaction of variables I and J. Again these measure deviations from the effects of the variables I and J alone, so that $\alpha_{1j}^{IJ} + \alpha_{2j}^{IJ} = 0$ for any j (and similarly for i);

α_{ijk}^{IJK} measures the 3-variable interaction of all variables. Again, $\alpha_{1jk}^{IJK} + \alpha_{2jk}^{IJK} = 0$ for each j and k.

Thus the complete description for all probabilities $p(i,j,k)$ requires 12 independent parameters:

1 general mean;

4 main effects (1 for I, 2 for J, and 1 for K);

5 2-variable interactions (2 for IJ, 1 for IK, and 2 for JK);

2 3-variable interactions α_{ijk}^{IJK} and α_{121}^{IJK} (all the others can be calculated from these, since their sum across any single variable is 0).

Since the data requires $M = 12$ cells, the system is completely determined.

The goal of the analysis may now be expressed in terms of the α-parameters. A structural model for the data is specified by assuming that a certain set of the α-parameters will be used to estimate the probabilities $p(i,j,k)$ (and thus that the main effects and interactions represented by these parameters will be used in the model) and that all other a-parameters will be assumed to be 0 (so that these main effects and interactions are assumed not to exist). If the model fits the data well, then it divides the effects and interactions into those which are important in describing the data and those which are not. Typical models are defined as follows:

(1) The only parameter used is the general mean μ. The model says that $\log p(i,j,k) = \mu$ for each $p(i,j,k)$. Thus none of the variables have any effect, and the observed data points can be assumed to be distributed randomly over the M cells.

(2) Only the general mean and the main effects are used. Now $\log p(i,j,k) = \mu + \alpha_i^I + \alpha_j^J + \alpha_k^K$: the logarithm of each cell probability is the sum of the general mean and a main effect from each variable. If both sides of the equation are exponentiated, it is seen that each cell probability is the product of four factors, one being the "general mean" and the other three being the main effects of each of the variables. This is the classical independence model.

(3) The general mean, all main effects, and a single 2-variable interaction effect are used. If the JK interaction is the one present, then $\log p(i,j,k) = \mu + \alpha_i^I + \alpha_j^J + \alpha_k^K + \alpha_{jk}^{JK}$. The interpretation of this model is that the variable I is independent of the various combinations of J and K.

(4) The general mean, all main effects, and two 2-variable interactions are used. For instance, $\log p(i,j,k) = \mu + \alpha_i^I + \alpha_j^J + \alpha_k^K + \alpha_{ij}^{IJ} + \alpha_{jk}^{JK}$. This is a conditional independence model: given variable J, variables I and K are independent.

The Algorithm

The algorithm CONTAB, which was used to implement the log-linear model, employs an iterative fitting scheme. Each model to be fit specifies that certain interactions among the variables are to be included. This means that the corresponding marginal totals of the fitted table must agree with the same marginal totals of the original table. For instance, in the 3-variable example considered previously, let $x(i,j,k)$ denote the observed count in cell (i,j,k). A marginal total is a sum across one or more variables. Thus the IJ marginal is $\sum_{k} x(i,j,k)$ and is written as $x(i,j,\cdot)$, where the dot replaces the variable which has been summed. Similarly, the K marginal is $\sum_{ij} x(i,j,k) = x(\cdot,\cdot,K)$. In example (3) above the general mean, all main effects, and the JK interaction are to be present in the model. This means that the grand total $N = x(\cdot,\cdot,\cdot)$, all of the 1-variable marginals $x(i,\cdot,\cdot)$, $x(\cdot,j,\cdot)$, and $x(\cdot,\cdot,k)$ and the 2-variable JK marginal $x(\cdot,j,k)$ are fixed by the model to have the same values as in the original data. The algorithm CONTAB begins with a uniform distribution of N observations over M cells, and successively adjusts the cell entries so that each of the marginals fixed by the model agrees with its specified value. Since the adjustment to fix one marginal will usually destroy the adjustment for the marginals previously considered, the process must cycle repeatedly through all marginals to be fitted. The algorithm does converge (the individual cell entries each approach a limiting value); the interactions terminate when a cycle through all marginals changes the cell entries by less than a specified amount.

Goodness-of-Fit

The goodness-of-fit of a model to the observed data is measured by the Information Statistic of Kullback (which is equivalent to the likelihood ratio statistic of [1]). Let $x(i,j,k)$ be the observed count in cell (i,j,k), and let $x^*(i,j,k)$ be the count predicted by a structural model. The Information Statistic, which measures the fit of the predicted distribution x^* to the observed distribution x, is

$$2I(x:x^*) = \sum_{ijk} x(i,j,k)\log \frac{x(i,j,k)}{x^*(i,j,k)} .$$

Under the null hypothesis that the model is correct, this statistic has a distribution which is asymptotically chi-square, with degrees of freedom equal to the number of cells in the table minus the number of independent parameters specified by the model. The value of this statistic can be compared to tables of the chi-square distribution to test goodness-of-fit.

The Information Statistic can be used to test a hierarchical collection of models. Suppose that x_1^* and x_2^* are the distributions predicted from two different structural models, and that all the α-parameters used in the model for x_1^* are also used in the model for x_2^*. Then the statistic $2I(x_2^*:x_1^*)$ satisfies the relation

$$2I(x:x_1^*) = 2I(x:x_2^*) + 2I(x_2^*:x_1^*).$$

The distribution of $2I(x_2^*:x_1^*)$ is also asymptotically chi-square, with degrees of freedom equal to the number of independent parameters specified by x_2^* which are not specificed by x_1^*. It may be used to test the hypothesis that the parameters specified by x_2^* but not by x_1^* are significant.

Odds

If all but one of the variables describing a data set may be regarded as independent (causes, or predictor variables), and the remaining variable is dependent (the result to be predicted) the preceding description may be specialized somewhat. Suppose, for example, that in the previous example the variable K is dependent and I and J are independent. The variable K has two categories. One convenient way to predict the variable K is to estimate the odds that $k = 1$ compared to $k = 2$ for a given combination of I and J: estimate the ratio $p(i,j,1)/p(i,j,2)$ for all i and j. From the general model it follows that the natural log of this ratio is given by:

$$\log \frac{p(i,j,1)}{p(i,j,2)} = (\alpha_1^K - \alpha_2^K) + (\alpha_{i1}^{IK} - \alpha_{i2}^{IK}) + (\alpha_{j1}^{JK} - \alpha_{j2}^{JK})$$

$$+ (\alpha_{ij1}^{IJK} - \alpha_{ij2}^{IJK})$$

$$= 2\alpha_1^K + 2\alpha_{i1}^{IK} + 2\alpha_{j1}^{JK} + 2\alpha_{ij1}^{IJK}.$$

Thus only the interactions involving K are needed to describe the odds.

REFERENCES

1. Bishop, Y.M.M., S.E. Fienberg and P.W. Holland: "Discrete Multivariate Analysis", MIT Press, Cambridge, 1975.
2. Kullback, S.: "The Information in Contingency Tables", U.S. Army Research Office Technical Report, 1974.
3. Gokhale, D.V. and S. Kullback: "The Information in Contingency Tables", mimeographed notes, George Washington University, 1976.
4. Solomon, H.: "Passive Restraint Systems and Accident Outcome", Final Report, DOR HS-4-00974, April 1975.

APPENDIX C

DATA ANALYSIS

Model Testing with CONTAB

The data for this study are classified in a 6-way contingency table by the variables:

Y: year - 1973 or 1974 (2 levels);

D: district - rural or residential/business (2 levels);

R: roadway - 2 lane or at least 4 lane (2 levels);

T: truck type - 3,4,5-axle semitrailers or double bottoms (4 levels);

W: weight - 10,000 lb intervals (7 levels);

F: fatality - is there a fatality to a car occupant or not? (2 levels).

The cross-classification produces 2.2.2.4.7.2 = 448 cells for the 14,574 observations. Note from the data listing in Appendix A that the observations are far from uniform. In particular, since double-bottom trucks weigh more than 20,000 lbs when empty, there are no observations for double-bottom trucks in the lowest weight class. These 16 cells (double-bottom trucks, 10-20,000 lbs, any year, any district, any roadway, any fatality) are structural zeros, and should not be considered part of the cross-classification. Thus the true number of cells is 448 - 16 = 432. The degrees of freedom of the various models' fit must be based on 432 cells.

A summary of the models' fit is presented in Table 5.

The numbered rows of the Table correspond to the 7 models tested. The hierarchical structure of the model is used to test the significance of each new interaction as it is added. These interaction tests are given in the unnumbered rows immediately following the rows for the appropriate models. For the rows of the Table corresponding to models, the columns are respectively:

— Interactions: a list of the interactions present in the model. The presence of an interaction implies that all lower-order interactions are also present. For example, the interaction DRF implies that the interactions D,R,F,DR,DF and RF are also present.

— IS: the Information Statistic $2I(x:x_i^*)$, where x is the original distribution of the observed data and x_i^* is the distribution predicted by the model;

TABLE 5

Summary of Models' Fit to BMCS Data

Model	Interactions	IS	df	p
1	YDRTW,F	633.8	215	0.000
2	YDRTW,DF	313.2	214	0.000
	DF	320.6	1	0.000
3	YDRTW,DF,RF	266.5	213	0.005
	RF	54.1	1	0.000
4	YDRTW,DRF	243.5	212	0.063
	DRF	23.0	1	0.000
5	YDRTW,DRF,YF	243.1	211	0.059
	YF	0.4	1	> 0.5
6	YDRTW,DRF,TF	236.8	209	0.087
	TF	6.7	3	0.080
7	YDRTW,DRF,WF	234.0	206	0.084
	WF	9 5	6	0.180

— df: the number of degrees of freedom of the model;

— p: the probability that a chi-square random variable with df degrees of freedom will exceed the value IS.

For the rows corresponding to interactions, the Interaction column gives the interaction whose significance is to be tested; IS gives the Information Statistic $2I(x_1^*:x_2^*)$, where x_2^* includes the interaction being tested and x_1^* does not; df is the difference in degrees of freedom between x_2^* and x_1^*, which is the number of degrees of freedom attributable to the interaction; and p is as before. The value of p tests the goodness-of-fit of the model or the interaction. Usual statistical practice suggests that models be rejected if p is less than 0.05 and that interactions be considered significant if p is less than 0.05.

Since F is the dependent variable of interest, the first model tests whether F is independent of the combination of all other variables. Model 1, if valid, would be the simplest explanation of the odds of fatality: it would say that these odds are constant regardless of all of the other variables. Model 1 fits poorly, with p = 0.000 (to three decimal places), and is rejected.

Model 2 adds a single interaction DF; it says that the odds of fatality depend only on the district D of the accident. This model is also rejected, with p = 0.000. However, the interaction DF is highly significant, with a p = 0.000: even though DF by itself is not sufficient to produce an accurate model, it explains a significant amount of the variation left in the original model. The interaction DF will thus be retained in the following models.

Model 3 adds to Model 2 a second interaction, RF. Model 3 says that the odds of fatality depend on both D and R, independently; if valid, it would allow the odds of fatality to be approximated by the product of a district factor and a roadway factor. This model also fails, with $p = 0.005$. The interaction RF is also significant with a $p = 0.000$.

Model 4 adds to Model 3 the 3-way interaction DRF. Since $p = 0.063$, Model 4 gives an acceptable fit to the data. The DRF interaction is clearly significant, with $p = 0.000$. This model says that the odds of fatality may be approximated by knowing only the disctrict-roadway combination of the accident. District and roadway do not act independently, but interact. Another way of interpreting the district-roadway interaction is that the 4 possible district and roadway combinations are really 4 levels of a single district-roadway variable.

Model 5 adds to Model 4 an interaction YF, to test whether the variable Y interacts significantly with F. It does not: since $p > 0.5$, the interaction YF is not significant. Further, the fit of Model 5 ($p = 0.59$) is worse than that of Model 4 ($p = 0.063$), since the small reduction in IS is more than offset by the reduction in df.

Model 6 adds to Model 4 the TF interaction. This interaction is not significant at the 0.05 level ($p = 0.08$). Model 6 does fit slightly better than Model 4 ($p = 0.087$ compared to $p = 0.063$). Thus the TF interaction improves the fit only slightly, and is far less important than the interactions DF,RF and DRF added previously.

Model 7 adds to Model 4 the WF interaction. The conclusions for WF are analogous to those for TF: WF is not significant ($p = 0.18$), WF improves the fit of Model 4 slightly ($p = 0.084$ for Model 7), but WF is even less important than TF, and far less important than DF,RF, and DRF.

Additional models were tested to see if the independent variable interaction YDRTW (and the lower-order interactions implied by it) are necessary to describe the data. It was found that the complete interaction YDRTW is necessary, so that no simplification is possible. Model 4 was thus selected as the simplest which gives a satisfactory approximation to the observed data.

Comment on the Model

Model 4 uses the two interactions YDRTW and DRF (and, of course, all lower-order interactions implied by these). This simplicity of form leads to simplicity both in interpretation of the model and in calculating cell estimates.

Estimation: Cell probabilities for the estimated distribution x^* prescribed by the model can be calculated directly as:

$$x^*(y,d,r,t,w,f) = \frac{x(y,d,r,t,w,\cdot) - x(\cdot,d,r,\cdot,\cdot,f)}{x(\cdot,d,r,\cdot,\cdot,\cdot)} ,$$

(with the notation for marginals of Appendix B). This is a model of conditional independence: F is independent of the combination of Y,T and W, given the DR combination.

Interpretation: From the formula immediately above it follows that the odds of a fatality are estimated by:

$$\frac{x^*(y,d,r,t,w,2)}{x^*(y,d,r,t,w,1)} = \frac{x(\cdot,d,r,\cdot,\cdot,2)}{x(\cdot,d,r,\cdot,\cdot,1)} .$$

Clearly the odds depend only on the DR combination, and are completely independent of Y,T and W. Thus, for example, *every* predicted odds of fatality for rural, 2-lane roads, for *any* combination of year, truck type, and weight, is the same. This common value is the quotient of the observed marginals $x(\cdot,d,r,\cdot,\cdot,2)/$ $x(\cdot,d,r,\cdot,\cdot,1)$ and equals 0.211. In the presentation of the data in Appendix A, the observed odds are calculated for each pair of data points. A crude estimate of the fit of this model is to compare these actual odds with the predicted values: all the odds on the first data page are predicted to be 0.211; on the second, 0.120; on the third, 0.059; and on the fourth, 0.061. Some variation from these predicted values is clearly present, usually in cases of low cell counts, but this variation does not exceed that which might reasonably be expected to occur by chance.

Regressions

Linear regressions were fit to the overall odds of fatality as a function of weight (for the data of all years, districts, roadways, and truck types) and to the odds for each of the 4 district-roadway combinations. These odds are displayed in Figure 1 of the text. The results of the regressions are summarized in Table 6.

TABLE 6

Summary of Regressions

Category	Slope	standard error of slope	R^2	t
rural 2-lane	0.000120	0.00023	0.05	0.51
rural 4-lane	0.000086	0.00037	0.01	0.23
residential/ business 2-lane	0.000613	0.00027	0.51	2.30
residential/ business 4-lane	0.000747	0.00034	0.50	2.22
overall	0.000968	0.00023	0.78	4.19

The regressions in Table 6 show that the odds for each rural category are fit by a straight line with a very slight positive slope, of magnitude less than its standard error. The small t-values show that the hypothesis that the slope is 0 (and thus that there is no relation between truck weight and fatality odds for these categories) cannot be rejected. The low correlation coefficients R^2 show that these two lines explain very little of the variation in the data. The odds for each residential/business category are fit by a line with a slight positive slope, of magnitude a bit greater than 2 standard errors. The t-values allow the hypothesis that the slope is 0 to be rejected (at a confidence level of about 0.02), and the correlation coefficients show that about half of the variation in each category is explained by the regression lines. The overall odds are fit with a line of a definite positive slope, of magnitude about 4 standard errors. The hypothesis that the slope is 0 is definitely rejected, and the correlation coefficient of 0.78 shows that over 3/4 of the total variation in the data is explained by the regression line.

CONTINGENCY TABLE ANALYSES OF HIGHWAY COMMERCIAL
VEHICLE ACCIDENTS

Lloyd L. Philipson, Parvis Rashti, Gerald A. Fleischer

University of Southern California,
University Park, Los Angeles, California 90007

I. INTRODUCTION

Under contract to the U.S. Department of Transportation's
National Highway Traffic Safety Administration, the Traffic Safety
Center of the University of Southern California conducted an analysis of approximately 3,000 commercial vehicle accidents occurring
in California. (Philipson, et al., 1978). Each of these accidents
was investigated by the California (State) Highway Patrol and ex-
tended accident reports prepared. These reports formed the data
base for the subsequent evaluation. Contingency table analysis
methodology was one of several evaluation techniques applied to
specific investigations of variable interrelationships, with the
objective of exposing possible accident occurrence or accident
severity "causations" among them.
The minimum discrimination information (MDI) approach was first
applied to the investigation of the interaction of the most significant variables (identified in an earlier linear regression analysis) in the occurrence of Jackknife-Before-Accident (JKBA). The
computer program CONTAB was employed in this investigation. A detailed discussion of this particular application is provided here
in order to illuminate the important features of the method. The
results of one other CONTAB analysis of particular interest, that
of the factors affecting accident severity, are then presented
more briefly.

2. ANALYSIS OF JACKKNIFE-BEFORE-ACCIDENT (JKBA) EVENTS

Data were developed in the study of the joint frequencies
the occurrence of JKBA associated with the number of drive axles

of a vehicle (DA), the occurrence of wheel lock-up (LU), and the condition of the road surface (RS). The different levels of each of these variables and respective indices used are defined in Table 1.

TABLE 1

JKBA Analysis Variables

Character-istic	Variable	Variable name in analysis description	Associ-ated Index	Levels	
				1	2
Jackknife-Before-Accident	JKBA	A	h	No	Yes
Drive Axles	DA	B	i	1	2
Lock-Up	LU	C	j	No	Yes
Road Surface	RS	D	k	Not Dry	Dry

Input

There are 16 cells for this $2 \times 2 \times 2 \times 2$ contingency table, with the observed (cross-tabulated) cell frequency values x_{hijk} given in Table 2†. The underlying cell probabilities are denoted by $p(hijk)$, and are subject to the constraint $p(\cdots) = 1$.

TABLE 2

Input Contingency Table for JKBA Analysis¶

JKBA(h)						RS(k)	
						1	2
	1	DA(i)	1	LU(j)	1	68	804
					2	6	55
			2	LU	1	73	732
					2	9	60
	2	DA	1	LU	1	24	31
					2	5	19
			2	LU	1	8	6
					2	3	3

---

† The cross-tabulated or joint frequencies were established from the accident data base by a standard cross-tabulation computer routine.

¶ Applicable sample (size 1,906) was all accidents involving vehicles incorporating semi- or full-trailers.

In Table 2, it is seen that an example of a one-way marginal is

$$x(1\cdots) = 68 + 804 + 6 + 55 + 73 + 732 + 9 \quad +60 = 1,807;$$

a two-way marginal,

$$x(11\cdot\cdot) = 68 + 804 + 6 + 55 = 933;$$

a three-way marginal,

$$x(111\cdot) = 68 + 804 + 872;$$

and a four-way marginal,

$$x(1122) = 55.$$

Computer Analysis

An object of this analysis is to be able to estimate or "predict" the probability of occurrence JKBA; i.e., $p(hijk)$ for $h = 2$; given any particular combination of levels of i,j,k. For example, for $i = 2$, $j = 1$, $k = 1$ (two drive axles, no lock-up, dry road), what are the estimates of $p(h211)$, the probabilities of occurrence ($h = 2$) or non-occurrence ($h = 1$) of JKBA? In the usual terminology of multiple regression, A is regarded as the dependent variable and B,C and D are independent variables.

The simplest prediction model assumes that A is independent of B,C,D; there is no predictive capacity in the levels of B,C,D as far as A is concerned. This is formulated in symbols in terms of an hypothesis as

$$H_1: \; p(hijk) = p(h\cdots).p(\cdot ijk).$$

The abstracted computer output in Table 3, provided by a run on the CONTAB program, describes this hypothesis as

JKBA

DA*LU*RS

corresponding to the two sets of marginals that are fitted to the data, the one-way A-marginal and the four three-way BCD marginals. Under H_1,

$$p*(hijk) = \frac{x*(h\cdots)x*(\cdot ijk)}{N},$$

where N is the total frequency (1,906) and $p*$ denotes the estimate of $p(hijk)$ deriving from the model defined by the hypothesis H_1.

TABLE 3

Summary of CONTAB Output for JKBA Analysis
Factors: JKBA*DA*LU*RS; Sample Size: 1,906

	Hypothesis	I.S.	I*	D.F.	PROB.
1.	JKBA DA*LU*RS	148.739	0.00	7	0.0000
2.	JKBA*RS DA*LU*RS	81.928	0.45	6	0.0000
3.	JKBA*DA DA*LU*RS	116.512	0.22	6	0.0000
4.	JKBA*LU DA*LU*RS	105.894	0.29	6	0.0000
5.	JKBA*RS JKBA*DA JKBA*LU	4.989	0.97	4	0.2884

Note that H_1 includes all the interactions among B,C,D, *but exclus-ive of* A. It is not necessary to assume *a priori* that any of these possible interactions are absent, as that will impose unnecessary structure on the model. Further, the fitting procedure will ad-just accordingly if there are in fact no interactions present among B,C,D. The interest is only in the associations of the de-pendent variable with the various possible individual and joint occurrences of the independent variables. There is no interest in the characterization of these latter joint occurrences by them-selves.

The design matrix for this analysis is presented in Table 4. There we see that H_1 uses columns 1 (L), 2 (τ_1^A), three-factor in-teraction BCD (column 15), and *therefore* all lower-order interac-tions, columns 9,10,11, and 3,4,5, involving only i,j,k, and B,C,D. Other τ-parameters are set equal to zero. The number of the latter equals the degrees of freedom (D.F.). In the present case, for H_1, D.F. $= 16 - 9 = 7$.

Now refer to Table 3 for the evaluation of the model given by H_1. The "minimum discrimination information" (MDI) statistic (I.S.), which essentially reflects the variation between the model's predictions and the data, is 148.739. Using the X^2 approximation to the MDI statistic (Gokhale 1978), this is highly significant; the probability of a value as great or greater occurring randomly |PROB| is zero to four decimal places. Thus H_1 is strongly reject-ed; its model is a very poor fit to the data, and therefore one or more of the variables B,C,D are seen to be needed for the predic-tion of $p(hijk)$.

If the value of the MDI statistic had been non-significant (with PROB > 0.05, say), the analysis would conclude that A is independent

TABLE 4

Design Matrix for 2 × 2 × 2 × 2 Contingency Table Illustration

Parameters or Column Number

Cell Index				1	2	3	4	5	6	7	8	9	10	11	12	13	14	15	16
h	i	j	k	L	τ_1^A	τ_1^B	τ_1^C	τ_1^D	τ_{11}^{AB}	τ_{11}^{AC}	τ_{11}^{AD}	τ_{11}^{BC}	τ_{11}^{BD}	τ_{11}^{CD}	τ_{111}^{ABC}	τ_{111}^{ABD}	τ_{111}^{ACD}	τ_{111}^{BCD}	τ_{1111}^{ABCD}
1	1	1	1	1	1	1	1	1	1	1	1	1	1	1	1	1	1	1	1
1	1	1	2	1	1	1	1	0	1	1	0	1	0	0	1	0	0	0	0
1	1	2	1	1	1	1	0	1	1	0	1	0	1	0	0	1	0	0	0
1	1	2	2	1	1	1	0	0	1	0	0	0	0	0	0	0	0	0	0
1	2	1	1	1	1	0	1	1	0	1	1	0	0	1	0	0	1	0	0
1	2	1	2	1	1	0	1	0	0	1	0	0	0	0	0	0	0	0	0
1	2	2	1	1	1	0	0	1	0	0	1	0	0	0	0	0	0	0	0
1	2	2	2	1	1	0	0	0	0	0	0	0	0	0	0	0	0	0	0
2	1	1	1	1	0	1	1	1	0	0	0	1	1	1	0	0	0	1	0
2	1	1	2	1	0	1	1	0	0	0	0	1	0	0	0	0	0	0	0
2	1	2	1	1	0	1	0	1	0	0	0	0	1	0	0	0	0	0	0
2	1	2	2	1	0	1	0	0	0	0	0	0	0	0	0	0	0	0	0
2	2	1	1	1	0	0	1	1	0	0	0	0	0	1	0	0	0	0	0
2	1	1	2	1	0	0	1	0	0	0	0	0	0	0	0	0	0	0	0
2	2	2	1	1	0	0	0	1	0	0	0	0	0	0	0	0	0	0	0
2	2	2	2	1	0	0	0	0	0	0	0	0	0	0	0	0	0	0	0

of B,C,D†; i.e., B,C,D have no predictive capacity as far as A is concerned. The quantity 148.739 (the value of the MDI statistic) can be looked upon as "unexplained variation" (by analogy with the analysis of variance) when the hypothesis of independence, H_1, is assumed. In a sequence of increasingly complex hypotheses to follow, the attempt is made to reduce this variation by including more and more interaction parameters in the model. The sequence is halted with as simple a model as possible, when a non-significant MDI statistic occurs, and a satisfactorily high percentage of variation in the data is explained.

In the output summarized in Table 3, the sequence derives from adding one or more interaction parameters in the following order:

H_2 includes τ_{11}^{AD} (column 8 of the design matrix);

H_3 includes τ_{11}^{AC} (column 7 of the design matrix);

H_4 includes τ_{11}^{AB} (column 6 of the design matirx)

H_5 includes all three of these parameters (columns 6,7 and 8).

In the output for H_2, the inclusion of τ_{11}^{AD} is reflected in the marginals fitted as

JKBA*RS   (A*D),

DA*LU*RS   (B*C*D).

The MDI statistic is 81.93 with 6 D.F., still highly significant (PROB of a value > I.S., given H_2, is still essentially zero) showing that the model provided by H_2 is not adequate and more interactions should be included to improve the model. The difference 148.74 - 81.93 = 66.81 with 7 - 6 = 1 D.F. measures the *effect* of inclusion of the interaction parameter τ_{11}^{AD}. A χ^2 table shows this is highly significant, indicating that the association of A with D should be incorporated in the model. (But the rejection of H_2 shows that the model has to include some other appropriate interactions as well). The quantity I* = 66.81/148.74 = 0.45 is the fraction of the variation in the "base" hypothesis, H_1, explained by including the interaction between A and D.

A similar analysis can now be made for the interaction parameter τ_{11}^{AC}, incorporated in the model under H_3. The MDI statistic is 116.51 with 6 D.F., so the model is still poor. The difference from H_1 is 32.23, however, so the interaction AC should be included in the model. (The difference is again highly significant.)

---

† Or more properly, that the hypothesis of this independence cannot be rejected with the data in hand.

The fraction of variation explained by the interaction AC alone is I* = 32.23/148.74 = 0.22.

Under H_4, the MDI statistic is 105.89, the difference is 42.85, and the explained variation is 0.29. So the conclusion is very similar to that for H_3: the model is still poor but the AB interaction is significant.

At this point, examination of the successive I* values enables the single most important predictor to be determined. All the interactions of JKBA with each single independent variable, A*D, A*C and A*B, are important enough to be included in the model. The single most important is the interaction with D (RS), the next is that with C (LU), and the last is that with B (DA).

The next step in the sequence of models to be exhibited is the inclusion of all three of the pairwise interactions of A with B,C and D simultaneously. (Simpler possibilities, each including one of the three possible sets of two or these, were also tried, but were not yet deemed satisfactory.) This hypothesis is indexed as H_5. The computer output for this model shows that the interactions included are:

JKBA*RS,

JKBA*LU,

JKBA*DA,

DA*LU*RS,

implying that over and above the parameters included in the base hypothesis (which fitted marginals JKBA and DA*LU*RS), interaction parameters τ_{11}^{AD} (corresponding to JKBA*RS), τ_{11}^{AC} (corresponding to JKBA*LU), and τ_{11}^{AB} (corresponding to JKBA*DA) are included.

The MDI statistic for this model is 4.99 with 4 D.F., which is not significant at the 28% level. More important, it accounts for $\frac{148.74 - 4.99}{148.74}$ = 0.97 of the unexplained variation in H_1. So H_5 can be regarded as an acceptable model. In this model all three-factor interactions can be taken to be negligible. Thus, also, a satisfactory prediction of the JKBA probability is decided to require all three separate cross-tabulations of data in the form JKBA*RS, JKBA*LU, and JKBA*DA only; but no three-way or four-way cross-tabulations involving JKBA.

The implication, then, is that JKBA causation depends, and its prediction therefore requires information, only on the interaction of JKBA with all three of the independent variables *individually*, and with RS more than with the others. However, no information is required on their joint occurrence among themselves and with JKBA. Thus a simple situation exists for the further study of the probability of occurrence of JKBA. Sample data need only be obtained on the joint occurrences of JKBA with the DA,LU, and RS levels separately, and the probability (or frequency) of JKBA can then be predicted simply by the model in which parameters τ_{111}^{ABC}, τ_{111}^{ABD}, τ_{111}^{ACD}, and τ_{1111}^{ABCD} (in the design matrix, Table 4) are set equal to zero.

A final note is in order on the sequence of hypothesis tests that have been made. It is evident that there is no fully *objective* criterion available for deciding where to stop the sequence and accept a model, if successive models exist that are all statistically non-rejectable and explain increasing fractions of initial variation. (Note that the ultimate possible model, which fits every input contingency table cell exactly, necessarily explains 100% of this variation. But it is clearly virtually certain to be spurious in its prediction of the "noise", as well as meaningful components of the observed data. Moreover, it is certainly neither simple nor useful.) A subjective decision must be made; the simplest possible model must be accepted that is both non-rejectable at a reasonable level of significance, and provides a reasonable explanation of the variation. Judgment, based on experience with both the general behavior of the accident parameters and with contingency table analysis, is an unavoidable requirement.

The Log-Linear Model

The log-linear model corresponding to the accepted hypothesis H_5 includes L and the first 10 τ's in the design matrix (Table 4). To estimate $\ln p(hijk)$ from a future set of data, only data sufficient to compute estimates of these 11 L and τ's should be required. For the present data set, it is of interest to see the L and τ values actually determined, and then to compare the table of joint frequencies, $x^*(hijk)$, *predicted* by the log-linear model they establish, with the original table of the *observed* $x(hijk)$. The $\tau(1)$ to $\tau(11)$ values (with $\tau(1)$ representing L), and x^* values are shown in Table 5.

Comparing x^* of Table 5 with x of Table 2 shows quite good agreement, as is expected. Table 6 gives a few examples†.

Use of Model to Predict JKBA Odds

The x^* values given in Table 5 for the accepted model can be used to develop the predicted (or "smoothed") odds of the occurrence of JKBA compared to its non-occurrence, under various conditions given by different combinations of the levels of the independent variables. The possible cases follow (recall Table 1 for the definitions of the levels of the variables):

---

† Confident intervals for the x^* values can also be generated assuming they are normally distributed (cf. Gokhale 1978).

TABLE 5

τ-Values and Resulting Predicted Joint Frequency Ta le
for JKBA Analysis

a. *Values of τ's under* H_5

$$\tau(1) = 2.768229$$
$$\tau(2) = 1.345972$$
$$\tau(3) = 0.583343$$
$$\tau(4) = -0.017297$$
$$\tau(5) = -1.270198$$
$$\tau(6) = 1.920491$$
$$\tau(7) = 2.912554$$
$$\tau(8) = 0.197128$$
$$\tau(9) = 0.729127$$
$$\tau(10) = 0.219917$$
$$\tau(11) = 0.612444$$

b. *Estimate of x under* H_5

Cell	Estimated x^*
1 1 1 1	68.99
1 1 1 2	801.65
1 1 2 1	3.58
1 1 2 2	58.78
1 2 1 1	75.28
1 2 1 2	731.07
1 2 2 1	8.15
1 2 2 2	59.49
2 1 1 1	23.01
2 1 1 2	33.35
2 1 2 1	7.42
2 1 2 2	15.22
2 2 1 1	5.72
2 2 1 2	6.93
2 2 2 1	3.85
2 2 2 2	3.51

TABLE 6

Comparison of x* with x in JKBA Analysis

Cell	x^* (Table 6)	x (Table 4)
1 1 1 11	68.99	68.00
1 2 2 2	59.49	60.00
2 1 1 1	23.01	24.00
2 2 2 2	3.51	3.00

(a) Dry road surface, one drive axle, no lockup:

$$\frac{x^*(2,1,1,2)}{x^*(1,1,1,2)} = \frac{33.35}{801.6} = 0.04;$$

(b) Wet road, one drive axle, no lockup:

$$\frac{x^*(2,1,1,1)}{x^*(1,1,1,1)} = \frac{23.00}{68.99} = 0.33;$$

(c) Dry road, one drive axle, lockup:

$$\frac{x^*(2,1,2,2)}{x^*(1,1,2,2)} = \frac{15.22}{58.78} = 0.26;$$

(d) Wet road, one drive axle, lockup:

$$\frac{x^*(2,1,2,1)}{x^*(1,1,2,1)} = \frac{7.42}{3.58} = 2.1;$$

(e) Dry road, two drive axles, no lockup:

$$\frac{x^*(2,2,1,2)}{x^*(1,2,1,2)} = \frac{6.93}{731.1} = 0.0095;$$

(f) Wet road, two drive axles, no lockup:

$$\frac{x^*(2,2,1,1)}{x^*(1,2,1,1)} = \frac{5.72}{75.28} = 0.076;$$

(g) Dry road, two drive axles, lockup:

$$\frac{x^*(2,2,2,2)}{x^*(1,2,2,2)} = \frac{3.51}{59.49} = 0.059;$$

(h) Wet road, two drive axles, lockup:

$$\frac{x^*(2,2,2,1)}{x^*(1,2,2,1)} = \frac{3.85}{8.15} = 0.47.$$

Clearly, a wet road surface is the primary factor in the odds of occurrence of JKBA (as the importance of the JKBA*RS interaction that was determined implies). About a factor of 8 increase from the dry road odds appears in all four pairs of cases comparing these two conditions for various combinations of the levels of number of drive axles and occurrence of lockup.

A secondary note is that, as would be expected, the presence of two drive axles significantly decreases the odds of JKBA. It is less to be expected (but is of course implied by the model given by H_5) that this decrease is given by the same proportion for either road surface when lockup occurs: on a dry road, by a factor of $0.26/0.059 = 4.4$, and on a wet road, by a factor of $2.1/0.47 = 4.5$. When lockup does not occur, the corresponding factors are $0.04/0.0095 = 4.2$ and $0.33/0.076 = 4.3$, still about the same as before. The conclusion of some significance is that two drive axles reduce the odds of occurrence of JKBA by about a factor of 4 under all conditions.

It is finally worth noting generally that a complete odds analysis enables the discrimination of the combinations of the levels of the independent variables that produce the lowest odds of a deleterious level of the dependent variable. To the extent that the independent variables' levels are controllable, counter-measures to the deleterious level's occurrence could then be defined by these combinations. For example, in the present case of JKBA, if it were reasonable to require two drive axles for certain vehicles that would not otherwise employ them, the odds of the occurrence of JKBA could be decreased.

II. A SECOND APPLICATION: CONTAB ANALYSIS OF INJURY SEVERITY

A contingency table analysis was also conducted that was as similar as possible to one previously reported on by Hedlund (1977). Hedlund was able to make use of a larger data base (derived from the nationwide Bureau of Motor Carrier Safety accident reports data) that also included certain factors not involved in the present study. Hedlund's variables were:

(a) Year, District, Roadway (two levels: two lanes, or at least four lanes);

(b) Truck Type (four levels: Three-, Four-, or Five-Axle Semi-Trailers, or Double-Bottoms);

(c) Weight (seven levels, 10,000 lb intervals);

(d) Fatality (two levels: of some involved car occupant, or not).

Certain of these variables also appeared, at least approximately, in the present data base. The smaller present sample size, however, necessitated some further aggregation of their levels. Thus, the present analysis treated:

(a) Road Type (two levels: conventional two-way, or freeway or expressway);

(b) Truck Type (two levels: semi-trailer, or full trailer [generally a double-bottom]);

(c) Weight (three levels: 10,000-25,000 lbs, 25,000-60,000 lbs, greater than 60,000 lbs);

(d) High-Severity Injury (two levels: for some car occupant, yes or no).

Here "High-Severity Injury" refers to the highest injury levels — major or fatality. This combination was necessitated by the relatively small number of fatalities occurring in the present data base.

The input Contingency Table for this analysis is shown in Table 7. A summary of the CONTAB output for the most important hypotheses tested is given in Table 8.

TABLE 7

Input Contingency Table
for Accident Severity Analysis

Injury Severity (S)	Road Type (RT)	Truck Type (TT)	Weight (W) 1	2	3
1	1	1	68	99	73
		2	20	86	87
	2	1	138	264	151
		2	68	166	158
2	1	1	12	15	14
		2	6	16	17
	2	1	7	34	22
		2	11	13	23

It is seen first that Injury Severity does depend upon the independent variables. Hypothesis 1 is quite strongly rejected, at

TABLE 8

*Summary of CONTAB Output
for Accident Severity Analysis.
Factors: S*RT*TT*W; Sample Size 1,568*

	Hypothesis	I.S.	I*	D.F.	PROB
1.	S RT*RR*W	21.158	0.00	11	0.0318
2.	S*RT RT*TT*W	12.800	0.40	10	0.2351
3.	S*TT RT*TT*W	20.622	0.03	10	0.0239
4.	S*W RT*RR*W	18.558	0.12	9	0.0292
5.	S*W S*TT S*W RT*TTW	10.337	0.51	7	0.1703
6.	S*RT*W S*TT*W RT*TT*W	2.583	0.88	3	0.4604
7.	S*RT*W S*TT*W S*RT*TT RT*TT*W	2.306	0.89	2	0.3156

about a 3% level of significance. It is next observed that Road
Type is the most important individual explanatory variable (Hypothesis 2); it alone explains 0.40 of the initial variation under
the first hypothesis. Moreover, this simple model is not rejected
at a 23.5% significance level. If all three two-way interactions
of Severity with Road Type, Truck Type, and Weight are incorporated
(Hypothesis 5), the explained variation is raised to 0.51, and the
model is not rejected at a 17% level of significance. Hypotheses
2 and 5 are thus both statistically very acceptable, but their models do not yet provide good predictions of the probabilities of the
injury severity levels.

If the output to this point in Table 8 is compared with Hedlund's
results, it appears that both studies agree that Road Type is the
most important single factor in interacting with Injury Severity
(or just Fatality, in Hedlund's case), other than District, not
presently treated. Hedlund's "model 4" incorporates this interaction and attains fair statistical acceptance at a 6.3% level of
significance. This model explains 0.62 of the initial variation.
In the present study, the Injury Severity/Road Type interaction

alone (Hypothesis 2) is seen to explain 0.40 of the initial variation, and provides an acceptable model statistically at a 23.5%
level of significance, as was previously noted. However, a much
better explanation of the variation, 88%, is provided by Hypothesis
6, also statistically acceptable (at a 46% level of significance),
which includes the two three-way interactions Severity/Road Type/
Weight and Severity/Truck Type/Weight.

The conclusion is reached that this last model is a satisfactory
one for predicting injury severity level. The third three-way interaction Severity/Road Type/Truck Type is not required; Hypothesis
7 shows this adds only 1% to the explained variation.

The final predicted x^* values deriving from the model given by
Hypothesis 6 are shown in Table 9. The original observed x-values
are also given. A good fit of x^* to x clearly exists.

Finally, it is of interest to develop certain odds predicted by
the present study's results. Four representative cases for the
odds of a high-severity injury compared to a low-severity one (or
no injury) are as follows (recall Table 7 for the definitions of
the variables and levels indices):

(a) Conventional two-way roads, lightweight (less than 25,000
lbs) semi-trailers:

$$\frac{x^*(2,1,1,1)}{x^*(1,1,1,1)} = \frac{10.6}{69.4} = 0.15;$$

(b) Freeways and expressways, lightweight semi-trailers:

$$\frac{x^*(2,2,1,1)}{x^*(1,2,1,1)} = \frac{8.4}{136.6} = 0.06;$$

(c) Conventional roads, heavyweight (greater than 60,000 lbs)
full-trailers:

$$\frac{x^*(2,1,2,3)}{x^*(1,1,2,3)} = \frac{17.6}{86.4} = 0.20;$$

(d) Freeways, heavyweight full-trailers:

$$\frac{x^*(2,2,2,3)}{x^*(1,2,2,3)} = \frac{22.4}{158.6} = 0.14.$$

It is thus clear that road type is indeed the dominant factor in
the odds of severe injuries, with conventional roads significantly
more involved with such injuries than freeways. The dominance is
more pronounced, however, for the lighter vehicles (a factor of
0.15/0.06 = 2.5) than for heavier vehicles (0.20/0.14 = 1.4).

TABLE 9

*Predicted Joint Frequency Table
from Accident Secerity Analysis*

Cell	Estimated x^*	Observed x
1111	69.44	68.00
1112	96.97	99.00
1113	73.58	73.00
1121	18.56	20.00
1122	88.03	86.00
1123	86.42	87.00
1211	136.57	138.00
1212	266.03	264.00
1213	150.42	151.00
1221	69.44	68.00
1222	163.97	166.00
1223	158.58	158.00
2111	10.56	12.00
2112	17.03	15.00
2113	13.42	14.00
2121	7.44	6.00
2122	13.97	16.00
2123	17.58	17.00
2211	8.44	7.00
2212	31.97	34.00
2213	22.58	22.00
2221	9.56	11.00
2222	15.03	13.00
2223	22.42	23.00

CONCLUSIONS

Computer-based contingency table analysis methods have been applied to a number of exemplary analyses of highway commercial vehicle accident factors. Two of these analyses have been discussed here. Using the method, relatively simple subsets of all the possible variable interactions have been shown to be satisfactorily explanatory of the dependent variable of interest in each case — the relative likelihood of occurrence of JKBA, and of high severity injuries.

It is concluded that such applications of the method should be developed further and more widely employed. However, to enable this, improved accident data reporting and data processing procedures are needed that will provide the greatest possible descriptive detail and accuracy. Recommendations for such improvements are included in the results of the study (Philipson, et al., 1978) from which this chapter derives.

REFERENCES

1. Bishop, Y.M.M., et al.: *Discrete Multivariate Analysis*, The MIT Press, Cambridge, Massachusetts (1975).
2. Fisher, Martin, et al.: *Computer Programs on Contingency Table Analysis*, Washington, D.C., George Washington University (n.d.).
3. Gokhale, D.V. and S. Kullback: *The Information in Contingency Tables*, Marcel Dekker, Inc., (1978).
4. Hedlund, J.: *The Severity of Large Truck Accidents*, Technical Note DOT HS-802322, National Highway Traffic Safety Administration, April 1977.
5. Ku, H.H. and S. Kullback: "Loglinear Models in Contingency Table Analysis", *The American Statistician*, 28(4), 115-122 (1974).
6. Philipson, L.L., et al.: *Statistical Analyses of Commercial Vehicle Accident Factors*, Report DOT HS-803418, prepared for the National Highway Traffic Safety Administration, University of Southern California, February 1978.

APPENDIX

APPENDIX

PERCENTILES OF THE CHI-SQUARE DISTRIBUTION

Degrees of Freedom	(Note: Use the complement for significance level)					df
	0.50	0.90	0.95	0.99	0.995	
1	0.4	2.7	3.8	6.6	7.9	1
2	1.4	4.6	6.0	9.2	10.6	2
3	2.4	6.2	7.8	11.3	12.8	3
4	3.4	7.8	9.5	13.3	14.9	4
5	4.4	9.2	11.1	15.1	16.7	5
6	5.3	10.6	12.6	16.8	18.5	6
7	6.3	12.0	14.1	18.5	20.3	7
8	7.3	13.4	15.5	20.1	22.0	8
9	8.3	14.7	16.9	21.7	23.6	9
10	9.3	16.0	18.3	23.2	25.2	10
11	10.3	17.3	19.7	24.7	26.8	11
12	11.3	18.6	21.0	26.2	28.3	12
13	12.3	19.8	22.4	27.7	29.8	13
14	13.3	21.1	23.7	29.1	31.3	14
15	14.3	22.3	25.0	30.6	32.8	15
16	15.3	23.5	26.3	32.0	34.3	16
17	16.3	24.8	27.6	33.4	35.7	17
18	17.3	26.0	28.9	34.8	37.2	18
19	18.3	27.2	30.1	36.2	38.6	19
20	19.3	28.4	31.4	37.6	40.0	20
21	20.3	29.6	32.7	38.9	41.4	21
22	21.3	30.8	33.9	40.3	42.8	22
23	22.3	32.0	35.2	41.6	44.2	23
24	23.3	33.2	36.4	43.0	45.6	24
25	24.3	34.4	37.7	44.3	46.9	25
30	29.3	40.3	43.8	50.9	53.7	30
35	34.3	46.1	49.8	57.3	60.3	35
40	39.3	51.8	55.8	63.7	66.8	40
60	59.3	74.4	79.1	88.4	92.0	60
80	79.3	96.6	101.9	112.3	116.3	80
99	98.3	117.4	123.2	134.6	139.0	99
120	130.4	140.2	146.6	159.0	163.6	120